T0189423

Studies in Computational Intelligence

Volume 593

Series editor

Janusz Kacprzyk, Polish Academy of Sciences, Warsaw, Poland
e-mail: kacprzyk@ibspan.waw.pl

About this Series

The series "Studies in Computational Intelligence" (SCI) publishes new developments and advances in the various areas of computational intelligence—quickly and with a high quality. The intent is to cover the theory, applications, and design methods of computational intelligence, as embedded in the fields of engineering, computer science, physics and life sciences, as well as the methodologies behind them. The series contains monographs, lecture notes and edited volumes in computational intelligence spanning the areas of neural networks, connectionist systems, genetic algorithms, evolutionary computation, artificial intelligence, cellular automata, self-organizing systems, soft computing, fuzzy systems, and hybrid intelligent systems. Of particular value to both the contributors and the readership are the short publication timeframe and the world-wide distribution, which enable both wide and rapid dissemination of research output.

More information about this series at http://www.springer.com/series/7092

Mohamed Chawki · Ashraf Darwish
Mohammad Ayoub Khan
Sapna Tyagi

Cybercrime, Digital Forensics and Jurisdiction

Mohamed Chawki
International Association of Cybercrime Prevention (AILCC)
Paris
France

Sapna Tyagi
Institute of Management Studies
Ghaziabad
India

Ashraf Darwish
Department of Computer Science, Faculty of Science
Helwan University
Cairo
Egypt

Mohammad Ayoub Khan
Department of Computer Science and Engineering, School of Engineering and Technology
Sharda University
Greater Noida
India

and

College of Computer Science and Engineering, Yanbu Branch
Taibah University
Medina
Kingdom of Saudi Arabia

ISSN 1860-949X ISSN 1860-9503 (electronic)
Studies in Computational Intelligence
ISBN 978-3-319-37726-1 ISBN 978-3-319-15150-2 (eBook)
DOI 10.1007/978-3-319-15150-2

Springer Cham Heidelberg New York Dordrecht London

Printed on acid-free paper

Springer International Publishing AG Switzerland is part of Springer Science+Business Media (www.springer.com)

Foreword

Well researched, hard-hitting, and visionary in application, *Cybercrime, Digital Forensics and Jurisdiction* is an essential reading for anyone involved with understanding and preventing cybercrime—business entities, private citizens, and government agencies. Because cyberspace is easily accessible, geographically unbounded, and largely unregulated it is an ideal vehicle for communication and business. Unfortunately, these same qualities also make cyberspace extremely vulnerable to a variety of untoward activities, to include a wide range of criminal activity. Ranging from malicious to felonious, cybercrime costs billions of dollars and can have serious consequences such as disrupting critical operations, causing loss of revenue and intellectual property, or loss of life. The authors of the book are world renowned subject-matter experts in cyber security issues, having taught, lectured, and written in this area across the globe for over 10 years. Drawing on scrupulous research in the field, the authors cover the full range of issues associated with cybercrime to include definitions, history, public policy, legal remedies, and civil liberty concerns. The authors have also placed all of these factors into an international framework, exploring the varied approaches from governments across the world. Covering in detail a variety of criminal activities conducted in cyberspace, the thrust of *Cybercrime* is firmly rooted in the law demonstrating that a viable strategy to confront cybercrime must be international in scope.

Jeffrey F. Addicott
Distinguished Professor of Law
Director, Center for Terrorism Law
St. Mary's University School of Law
One Camino Santa Maria
San Antonio, Texas

Preface

In the opinion of some commentators, crime has always been with us and computers are simply another tool criminals use to commit their offences. Accordingly, in one of the first books to place computers and law in the proper context, Colin Tapper asks the rhetorical question why there is "any more need for a book on the law of computers than there is for on the law of typewriters or tuning forks" (van der Merwe 2008, p. 61).

Tapper's answer boils down to the fact that computers were playing an increasingly important role in the community, that they were qualitatively different from anything else that had come before, and that traditional legal principles were inadequate to deal with the legal problems caused by computers (Ibid). Well into the twenty–first century these arguments seem more valid than ever. Not only has "the computer" been expanded into a wide-ranging and dynamic new concept called "information and telecommunications technology" (ICT), but the law is still desperately trying to keep up with this rapidly moving target (Ibid).

In the field of fraud by computer manipulation, many criminal law systems "de lege lata" face considerable difficulties in applying their traditional criminal provisions (Sieber 1998, Online). The statutory definitions of theft, larceny, and embezzlement in many legal systems are not applicable if the perpetrator appropriates deposit money; in many countries these provisions also cause difficulties as far as manipulations of cash dispensers are concerned (Ibid). The statutory definitions of fraud in many legal systems demand that a person be deceived. They cannot be used when a computer is "cheated" (Ibid).

At the same time that society is becoming more fragile, potential threats are expanding, diversifying, and becoming more opaque. The capability to launch cyberattacks that can exploit these strategic vulnerabilities is "shifting to the edge," that is, technology is empowering networked groups and individuals to directly threaten nation state security (and, thus, global stability) by putting critical infrastructure at risk (Brown et al. 2009, p. 671). Examples include:

- The "Millennium Bug" or Y2K program of 1997–1999, which led to a complete inventory of computing inside large organizations, often for the first time since the deployment of the enterprise Personal Computer (PC) in the mid-1980s;
- Denial of Service (DoS) attacks, beginning in 2001 against Yahoo! and eBay;
- Business continuity planning in the wake of the attacks in September 11, 2001;
- Corporate responses to the increasing financial returns for attackers (for example the growth of "phishing" and the 2004–2005 cyber-extortion cases against gambling websites);
- Cyberattacks against Estonia's government websites, 2007;
- 2010 FIFA World Cup Fraud;
- "Hacktivisit" supporters of Julian Assange allegedly shutting down MasterCard website in December 2010;
- U.S. defense firm Lockheed Martin struck by cyberattack in May 2011.

Intended Audience and Methodological Orientation

The book's main academic reference point is the intersection between law and criminology, particularly as it relates to the "law in action." Many harmful internet behaviors that raise public concern do not necessarily fall neatly within the criminal or civil codes. Furthermore, how they are currently resolved is framed increasingly by the broader discourse of public safety, as well as specific law enforcement debates. The book's intended audiences are advanced undergraduates and graduate students as well the various professional communities involved in cybercrime-related policy-making or practice.

This book contains a discussion about cybercrimes and the management issues they raise. The narrative is constructed from a grounded analysis of events and draws upon a range of different sources, as the distributed (as opposed to centralized) organization of the Internet creates multiple flows of often conflicting information about cybercrime offending and victimization. Networked information has the tendency to flow quite freely in all directions and so evade the editorial verification and control of dissemination which characterized previous media ages. Some of these information sources are the product of impartial and reliable news reporting, others are not. The latter may be the product of vested interest, or the results of academic research or surveys conducted by the many governmental and nongovernmental agencies that have an interest in most things "cyber".

Whenever possible, original and independent sources of information have been sought to ensure the reliability of information and in order to either (i) confirm a particular trend in two or more independent datasets or (ii) record the occurrence of an event. Adopting such an approach has helped to counter polemical accounts. Particular care was taken to avoid following purely sensationalized news streams, where one article reporting a sensational news event tends to spawn a chain of others.

The resulting analysis therefore takes a multidisciplinary approach to respect the many voices of crime. The digital realist perspective adopted here is not, however, to be confused with the artistic theory of pseudo-realism which carries the same name, or indeed, debates about left-realism. Rather, it originates in the work of many scholars who have been adapted here to contextualize cybercrime. Essentially, it is a multiple discourse approach which recognizes that cybercrime, like ordinary crime, is a form of behavior that is mediated by technology but also by social and legal values and economic drivers. The interrelation between these four influences not only shapes the digital architecture of criminal opportunity, but also provides some directions for resolving the same harms.

What follows, then, is a systematic enquiry based on available knowledge, literature, and research findings. It is also informed by my own research projects into cybercrime and cyber law conducted between 2003 and 2011.

What Is Covered?

This book is divided into five parts, each covering specific topics relating to cybercrime: What follows is a brief summary:

Part I: Fundamentals of Cybercrime
This part begins with an overview of cybercrime, the difference between cybercrime and similar offences, and the scale of the problem in the US, the UK, and the Middle East. This will be followed by a case study of several cybercriminals, as well as an illustration of the major challenges faced in fighting against this threat. The section concludes with an in-depth explanation of the future of Internet crime and punishment.

Part II: Computer System as Target
Part II of the book addresses the topic of computer hacking, by first covering the criminal statutes aimed at protecting computers in the US and the UK, and thereafter exploring types of malicious code, and the threats posed by viruses, worms, and Trojan horses. It concludes with an analysis of current threats and legislative approaches in the US, and the Council of Europe Convention on Cybercrime.

Part III: Computer System as Tool
Part III explains how the Internet is used as a tool and medium by cybercriminals. This includes the problem of phishing and the use of stolen financial information by phishers. The final chapter covers sexual harassment, providing readers with case studies to illustrate the prevalence of sexual harassment in cyberspace, enforcement problems, and the regulation of this threat in the US and the UK.

Part IV: Content—Related Offences
Part IV focuses on the role of the Internet in promoting child sexual abuse. This includes production of pornographic content, distribution and sharing, and the consumption of pornographic content. Criminal offences against children such as cybering, grooming, age play, and exposure to obscenity are also covered. The section ends with some strategies to minimize risks to youth when they access the Internet.

Part V: Privacy, Security, and Crime Control
Part V seeks to explore and analyze anonymity in cyberspace. It shows that the balance between privacy of Internet users and security of nations can be achieved. Furthermore, it focuses how anonymity can be achieved on the Internet and why it is an essential tool for free speech. This section will also describe national and regional strategies in Europe to fight against cybercrime. This section covers the Nigerian 419 scam, the operation of the scheme, the nature and extent of the problem, and an evaluation of the current situation in Nigeria.

Mohamed Chawki
International Association of Cybercrime
Prevention (AILCC)
Paris
France

Ashraf Darwish
Mohammad Ayoub Khan
Sapna Tyagi

References

I. Brown, L. Edwards, C. Marsden, in *Information Security and Cybercrime*, ed. by L. Edwards, C. Waelde. Law and the Internet (Hart, Oxford, 2009)
U. Sieber et al., *The Legal Aspects of Computer Crime and Security: A Comparative Analysis with Suggestions for Future International Action* (Elsevier, New York, 1998)
D. van der Merwe et al., *Information and Communications Technology Law* (LexisNexis, Durban, 2008)

Contents

Part II Computer System as Target

Part III Computer System as Tool

Part V Privacy, Security and Crime Control

Authors' Biography

Dr. Mohamed Chawki is Senior Judge in Egypt, the Founder Chairman of the International Association of Cybercrime Prevention (AILCC) in Paris, and the Founder Co-director of the African Center for Cyberlaw and Cybercrime Prevention (ACCP) in Kampala, Uganda. He received his Ph.D. in Law from the University of Lyon III in France, which was followed by a 3-year postdoctoral fellowship at the University of Aix-Marseille III. Dr. Chawki served as a legal consultant to the Chairman of the Egyptian Financial Supervisory Authority, where he has handled matters involving issues of public, private, and Egyptian capital market laws and is a former Legal Counselor to the Chairman of the Capital Market Authority (CMA) and to the Minister of Military Production in Egypt.
URL: http://www.cybercrime-fr.org
e-mail: chawki@cybercrime-fr.org

Dr. Ashraf Darwish received the Ph.D. in computer science from Saint Petersburg State University, Russian Federation in 2006 and is currently an assistant professor at the Faculty of Science, Helwan University, Egypt. Dr. Darwish teaches artificial intelligence, information security, data and web mining, intelligent computing, image processing (in particular, image retrieval and medical imaging), modeling and simulation, intelligent environment, and body sensor networking.

Dr. Darwish is the author of many scientific publications and his publications include papers, abstracts, and book chapters by Springer and IGI publishers. He keeps in touch with his mathematical background through his research. His research, teaching, and consulting mainly focuses on artificial intelligence, information security, data and web mining, intelligent computing, image processing, modeling and simulation, intelligent environment, body sensor networks, and theoretical foundations of computer science.
e-mail: modarwish@yahoo.com

Dr. Mohammad Ayoub Khan is working as a Professor in Taibah University, KSA and Department of Computer Science and Engineering, School of Engineering and Technology, Sharda University, Gr. Noida, India with cyber security protocol, hardware mechanism for security, Radio Frequency Identification system,

microcircuit design, and signal processing, NFC, front end VLSI (Electronic Design Automation, Circuit optimization, Timing Analysis), Placement and Routing in Network-on-Chip, Real Time and Embedded Systems. He has more than 11 years of experience in his research area. He has published more than 60 research papers in the reputed journals and international IEEE conferences. He is contributing to the research community by various volunteer activities. He has served as conference chair in various reputed IEEE/Springer international conferences. He is a senior member of professional bodies of IEEE, ACM, ISTE, and EURASIP societies.
e-mail: ayoub@ieee.org

Sapna Tyagi Her research interests include radio frequency identification system, Data warehousing, Knowledge discovery, data analysis, decision support, and the automatic extraction of knowledge from RFID data, cybercrime, and network security. She has more than seven years of experience in her research area. She is contributing to the research community by various volunteer activities (reviewing, editing). She is a member of IEEE, IACSIT, and IAENG.

Part I
Fundamentals of Cybercrime

Chapter 1
Cybercrime: Introduction, Motivation and Methods

1.1 Introduction

Advancements in modern technology have helped countries to develop and expand their communication networks, enabling faster and easier networking and information exchange. In less than two decades, the internet has grown from a curiosity to an essential element of modern life for millions. In addition to the socio-economic benefits, there is no doubt about computer technology and the internet that enhances the capabilities of human interaction. But somewhere the growth of global connectivity is inherent to cybercrime. Table 1.1 illustrates some interesting facts about the usage of ICT.

1.1.1 Definition

Computer-related crime or "cybercrime" or "e-crime" or "digital technology crime" is a long-established phenomenon, but the growth of global connectivity is inseparably tied to the development of contemporary cybercrime. Any criminal activity that involves a computer either as an instrument, target or a means for perpetuating further crimes comes within the ambit of cybercrime. A generalized definition of cybercrime may be "*unlawful acts wherein the computer is either a tool or target or both*".

The proliferation of digital technology and the convergence of computing and communication devices have transformed the way in which we socialise and do business. While overwhelmingly positive, there has also been a dark side to these developments. Crime follows opportunity; virtually every advance has been accompanied by a corresponding niche to be exploited for criminal purposes. "Cybercrime" has been used to describe a wide range of offences, including offences against computer data and systems (such as "hacking"), computer-related forgery

M. Chawki et al., *Cybercrime, Digital Forensics and Jurisdiction*,
Studies in Computational Intelligence 593, DOI 10.1007/978-3-319-15150-2_1

Table 1.1 Global connectivity scenario

In 2014, more than one third of the world's total population had access to the internet
Over 60 % of all internet users are in developing countries, with 45 % of all internet users below the age of 25 years
It is estimated that mobile broadband subscriptions will approach 70 % of the world's total population by 2017
In the future hyper-connected society, it is hard to imagine a 'computer crime', and perhaps any crime, that does not involve electronic evidence linked with internet protocol (IP) connectivity
There are nearly 2 billion internet users and over 5 billion mobile phone connections worldwide
Every day, 294 billion emails and 5 billion phone messages are exchanged

and fraud (such as "phishing"), content offences (such as disseminating child pornography), and copyright offences (such as the dissemination of pirated content).

The magic of digital cameras and sharing photos on the Internet is exploited by child pornographers. The convenience of electronic banking and online sales provides fertile ground for fraud. Electronic communication such as email and SMS may be used to stalk and harass. The ease with which digital media may be shared has led to an explosion in copyright infringement. Our increasing dependence on computers and digital networks makes the technology itself a tempting target; either for the gaining of information or as a means of causing disruption and damage. The idea of a separate category of 'computer crime' arose at about the same time that computers became more mainstream.

Generally speaking, computers play four roles in crimes: They serve as objects, subjects, tools, and symbols (Parker 1998). Computers are the objects of crime when they are sabotaged or stolen. There are numerous cases of computers being shot, blown up, burned, beaten with blunt instruments, kicked, crushed and contaminated (Ibid). The damage may be intentional, as in the case of an irate taxpayer who shot a computer four times through the window of the local tax office; or unintentional, as in the case of a couple who engage in sexual intercourse while sitting on computer sabotage and destroy information, or at least make it unavailable. Computers play the role of subjects when they are the environment in which technologies commit crimes. Computer virus attacks fall into this category. When automated crimes take place, computers will be the subjects of attacks. The third role of computers in crime is as tools, enabling criminals to produce false information or plan and control crimes. Finally, computers are also used as symbols to deceive victims. In a $50 million securities-investment fraud case in Florida, a stock broker deceived his victims by falsely claiming that he possessed a giant computer and secret software to engage in high-profit arbitrage. In reality, the man had only a desktop computer that he used to print false investment statements. He deceived new investors by paying false profits to early investors with money invested by the new ones (Parker 1998).

In the United States, police departments are establishing computer crime units, and cybercrime makes up a large proportion of the offences investigated by these divisions. The National Cybercrime training Partnership (NCTP) encompasses local, state, and federal law enforcement agencies in the United States.[1] The International Association of Chiefs of Police (IACP) hosts an annual Law Enforcement Information Management training conference that focuses on IT security and cybercrime.[2] The European Union has created a body called the forum on Cybercrime, and a number of European states have signed the Council of Europe's Convention on Cybercrime treaty, which seeks to standardize European laws concerning cybercrime. From this perspective, each organization and policy maker has their own ideas of what cybercrime is and isn't. These definitions may vary considerably. To effectively discuss cybercrime in this part, however, we need a working definition. Toward that end, we start with a board, general definition, before moving towards a more specific one.

When speaking about cybercrime, we usually speak about two major categories of offence: in the first, a computer connected to a network is the target of the offence; this is the case of attacks on network confidentiality, integrity and/or availability.[3] The other category consists of traditional offences such as theft, fraud, and forgery which are committed with the assistance of/or by means of computers connected to a network, computer networks and related information and communications technology. Cybercrime ranges from computer fraud, theft and forgery to infringements of privacy, the propagation of harmful content, the falsification of prostitution, and organized crime. In many instances, specific pieces of legislation contain definitions of terms. However legislators don't always do a good job of defining terms (Shinder 2002, p. 6). Sometimes they don't define them at all, leaving it up to law enforcement agencies to guess, until the courts ultimately make a decision (Ibid). One of the biggest criticisms to the definition of computer crime conducted by the U.S Department of Justice (DOJ) is of its overly broad concept. The (DOJ) defines computer crime as "any violation of criminal law that involved the knowledge of computer technology for its perpetration, investigation, or prosecution". Under this definition, virtually any crime could be classified as a computer crime, simply because a detective searched a computer database as part of the investigation.

One of the factors that make a hard-and-fast definition of cybercrime difficult is the jurisdictional dilemma. Laws in different jurisdictions define terms differently, and it is important for law enforcement officers who investigate crimes, as well as network administrators who want to become involved in prosecuting cybercrime that are committed against networks, to become familiar with the applicable laws (Ibid).

---

[1] *See* NCTP http://www.nctp.org.

[2] Ibid.

[3] Ibid.

One of the major problems with adequately defining cybercrime is the lack of concrete statistical data on these offences. As the reporting of crime is voluntary, the figures are almost certainly much lower than the actual occurrence of networked-related crime.

In many cases, crimes that legislators would call cybercrimes are just the 'same old stuff', except that a computer network is somehow involved. The computer network gives criminals a new way to commit existing crimes. Statutes that prohibit these acts can be applied to people who use a computer to commit them as well as to those who commit them without the use of a computer or network (Parker 1998, p. 114).

In other cases, the crime is unique and came into existence with the advent of the network. Hacking into computer systems is an example; while it might be linked to breaking and entering a home or business building, the elements that comprise unauthorized computer access and physical breaking and entering are different.

Most US states have laws pertaining to computer crime. These statutes are generally enforced by state and local police and might contain their own definitions of terms. The Texas Penal Code's Computer Crime section, defines Breach of Computer Security as "A person commits an offence if the person knowingly accesses a computer, computer network, or computer system without the effective consent of the owner".

California Penal Code, on the other hand, defines a list of eight acts that constitute computer crime, including: altering, damaging, deleting, or otherwise using computer data to execute a scheme to defraud; deceiving, extorting, or wrongfully controlling or obtaining money, property, or data using computer services without permission; disrupting computer services; assisting another in unlawfully accessing a computer; or introducing contaminates into a system or network. Thus, the definition of cybercrime under state law differs, depending on the state. Perhaps we should look to international organizations to provide a standard definition of cybercrime.

At the Tenth United Nations Congress on the Prevention of Crime and Treatment of Offenders, in a workshop devoted to the issues of crimes related to computer networks, cybercrime was broken into two categories and defined as:

(a) in a narrow sense: any illegal behaviour directed by means of electronic operations that targets the security of computer systems and the data processed by them.
(b) in a border sense: any illegal behaviour committed by means of, or in relation to, a computer system or network, including such crimes as illegal possession and offering or disturbing information by means of a computer system or network.

These definitions, although not completely definitive, do give us a good starting point on what we mean by *cybercrime*. Cybercrime, according to these definitions, involves computers and networks. In cybercrime, the "cyber" component usually refers to perpetrating qualitatively new offences enabled by information technology or integrating cyberspace into more traditional activities. Having defined the concept of cybercrime, it becomes necessary to compare it with traditional crime.

This involves examination of its characteristics, what makes it vulnerable to being manipulated, and the reports that have been conducted on its incidence and the damage it inflicts.

1.1.2 Contemporary Cybercrime

Cybercrime is one of the fastest growing areas of crime. More and more criminals are exploiting the speed, convenience and anonymity that modern technologies offer in order to commit a diverse range of criminal activities. In past, cybercrime has been committed by individuals or small groups of individuals. However, we are now seeing an emerging trend with traditional organized crime syndicates and criminally minded technology professionals working together and pooling their resources and expertise.

1.1.3 Terrestrial Crimes Versus Cybercrimes

The act of defining crime is often, but not always, a step toward controlling it. That is, the ostensible purpose of defining illegal behaviours as criminal is to make them liable to public prosecution and punishment (Crutchfield 2000, p. 7). Historically, "crime" was addressed at the local, community level of government (Hitchens 2003). Crime was small-scale, consisting of illegal acts committed by some persons that were directed against a victim. "Crimes", which were consistent across societies; fell into routinized, clearly-defined categories that reflected the basic categories of anti-social motivation, Crime was murder, robbery, and rape (Balckstone 1979).

Crime was also personal, if the victim and the offender did not know each other; they were likely to share community ties that put offences into a manageable, knowable context (Goodman and Brenner 2000, p. 151). This principle did not only facilitate the process of apprehending offenders—who stood a good chance of being identified by the victim or by reputation—but also gave citizens the illusion of security, the conceit that they could avoid being victimized if they avoided some activities or certain associations (Ibid). Law enforcement officers dealt with this type of crime because its parochial character meant investigations were limited in scope and because the incidence of crime stood in relatively modest proportion to the size of the local populace. Law enforcement's effectiveness in this regard contributed to a popular perception that social order was being maintained and that crime would not go unpunished (Ibid).

The development of ICTs in urbanization and in geographical mobility undermined this model to some extent, although it continued to function effectively for the most part. Legislators quickly adapted to the fact that ICTs could be used to commit fraud and to harass others. Because they modified their substantive criminal law to encompass these activities, the old model still functions effectively for traditional real world crime.

Unlike traditional crime, however, cybercrime is a global crime. As a European Report explains:

> computer-related crimes are committed across cyberspace and don't stop at the conventional state-borders. They can be perpetrated from anywhere and against any computer user in the world.

In order to understand the sea of change ICTs introduce into criminal activity, it is important to consider a hypothetical: one can analogize a denial of service attack to using the telephone to shut down a supermarket business, by calling the business' telephone number repeatedly, persistently, without remorse, and thereby preventing any other callers from getting through to place their orders. On such a base, the vector of cyberspace lets someone carry out an attack such as this easily, and with very little risk of apprehension. It is so easy, in fact, that a 13 year-old hacker used a denial of service attack to shut down a computer company. Furthermore, in addition to the increased scale of criminal activity that cybercrime offers, it also has a tendency to evade traditional offence categories. While some of its categories consist of using ICTs to commit traditional crimes, it also manifests itself as new varieties of activity that cannot be prosecuted using traditional offence categories.

The dissemination of the "Love Bug" virus illustrates this. Virus experts quickly traced this virus to the Philippines. Using Information supplied by an internet service provider, agents from the Philippines' National Bureau of Investigation and from the FBI identified individuals suspected of creating and disseminating the "Love Bug". However, they ran into problems with their investigation: the Philippines had no cybercrime laws, so creating and disseminating a virus was not a crime. Law enforcement officers had no hard time convincing a magistrate to issue a warrant to search the suspects' apartment. Later on, moreover, the suspected author of the virus could not be prosecuted under the repertoire of offences defined by the Philippines criminal code.

Cybercrime's ability to morph into new and different forms of antisocial activity that evade the reach of existing penal law creates challenges for legislations around the world. Criminals have the ability of exploiting gaps in their own country's penal law in order to victimize their fellow citizens with impunity. Additionally, cybercriminals can exploit gaps in penal laws of other countries in order to victimize those nation's citizens, and others. As the "Love Bug" episode demonstrated, cybercrime is global crime.

1.1.4 Cybercrime and White Collar Crime

The definition of white collar crime has been an enduring topic of debate over the past century. Some are of the opinion that white collar crime is a "social rather than a legal concept, one invented not by lawyers but by social scientists". There is no specific offence or group of offences that can be identified as white collar crime. As such, white collar crime is a concept similar to cybercrime in definitional difficulties (Smith et al. 2004, p. 10).

The traditional definition of white collar crime focused on crimes committed by persons of high status and social repute in the course of their occupation. Included in this definition were crimes committed by company officers, public servants, and professionals such as doctors and lawyers. The original emphasis was on economic crime, although over time, white collar crime has come to include any acts of occupational deviance involving a breach of the law or ethical principles. As such, it has been suggested that white collar crime now includes almost any form of illegality other than conventional street crimes (Ibid).

Technological developments over the past decade have created further complexities surrounding the types of persons able to commit white collar crime. The perpetration of an online fraud for example, might just as easily be a self-taught teenager using personal computer at home, as an educated professional in the workplace.

A simple categorisation distinguishes crimes committed by specified types of offenders from crimes perpetrated in specified ways. The essence of white collar crime, however, remains rooted in abuse of power and breach of trust, and usually involves the pursuit of financial gain as a motive (Ibid).

Clearly, not all white collar crimes involve the use of digital technologies, although in recent times the vast majority have done. Examples of those which do not include acts of violence committed in the workplace, such as the sexual assault of patients by doctors, and some environmental crimes, such as pollution (although even the later can be committed electronically).

The category of fraud or financial crimes of dishonesty intersects with white collar crime, economic crime, and cybercrime. Overlying these concepts are the categories of property crime and corporate crime. Property crime is sometimes used synonymously with economic crime, although trespass, for example, or acts of vandalism would not be economic, but nonetheless clearly property-related.

Problems also arise in relation to the types of property protected by criminal law. Notably, at least until recently, information has usually been regarded as being outside the scope of criminal prosecutions, dealt with instead by a range of intellectual property regimes such as copyright, patents, designs and protection of confidential information (Ibid).

1.2 The Scale of the Problem and Reasons for the Growth of Cybercrime

Knowing how much crime is committed might help us decide on how much to spend on security. Estimates by security experts of annual losses from computer crime range from $555 million to more than $13 billion (Parker 1998, p. 10), but there are actually no valid statistics on the losses from this type of crime, because no one knows how many cases go unreported. Even when the victims of computer crimes are aware of the crimes, they are usually relocated to report their losses, especially if those losses can be easily hidden (UNESCO 2000). Victims can lose more from reporting crimes than they lose from the crimes themselves. Embarrassment, key staff diverted to prepare evidence and testify, legal fees, increased insurance premiums, and exposure of vulnerabilities and security failures can all result from reporting computer crime incidents (Parker 1998, p. 10).

As every aspect of commerce and communication has been changed by the Internet, crime has evolved to profit from the millions of potential victims connected to one global network. There are various reasons for the growth of cybercrime. First of all, the technology for cybercrime has become more easily accessible. Software tools can be purchased online that allow the user to locate open ports or overcome password protection. These tools allow a much wider range of people to become offenders, not just those with a special gift for computing. Cybercrime's place is growing due to the exponential connectivity, increased subject knowledge, cultural awareness, and programmable onboard electronics, which increase the number of potential targets. Compared to other crimes and offenses, it generally requires a smaller investment and can be carried out in various locations, without any geographical constraints, with no consideration to borders.

1.2.1 United States

The results of national surveys bear out the picture that cybercrime is consistently and dramatically on the increase. One of the famous cited national surveys for the United States is the "Computer Crime and Security Survey" conducted by the Computer Security Institute, with the participation of the San Francisco branch of the Federal Bureau of Investigation's Computer Intrusion Squad.

Key findings from the 2010/2011 CSI Computer Crime and Security Survey include:

- Almost half of the respondents had experienced a security incident, with 45.6 % of them reporting they had been subject of at least one targeted attack.
- Malware infection continues to be the most commonly experienced attack.
- Fewer financial frauds were reported than in previous years, with only 8.7 % saying they had seen this type of incident.
- Tools that improve visibility into networks, web applications, and endpoints were ranked among the highest on information security and information technology managers' 'wish lists', including better log management, security information and event management, security data visualization, and security dashboards.

The survey was conducted from July 2009 through June 2010 and respondents included US security practitioners from the private and public sectors. In 2010, IC3 received the second highest number of complaints since its inception, whilst also reaching a major milestone this year when it received its two-millionth complaint. On average, IC3 receives and processes 25,000 complaints per month.

The most common victim complaints in 2010 were the non-delivery of payment/merchandise, scams impersonating the FBI and identity theft. Victims of these crimes reported losing hundreds of millions of dollars.

The 2010 Internet Crime Report demonstrates how pervasive online crime has become, affecting people in all demographic groups. The report provides specific details about various crimes, their victims and the perpetrators. It also shows how IC3 continually adapts its methods to meet the needs of the public and law enforcement.

1.2.2 United Kingdom

In the UK, the Office of Cyber Security and Information Assurance (OCSIA) works in partnership with Detica to look more closely at the cost of cybercrime and to gain a better appreciation of the costs to the UK economy of Intellectual Property (IP) theft and industrial espionage.

A study conducted considered three types of cybercrime that impact on individual citizens: identity theft; online scams and scareware.

The impact of identity theft was estimated in two ways, based on information published by CIFAS, in particular:

- The number of reported incidents was multiplied by the average cost of an incident and a further estimate made for the level of under-reporting;
- The number of UK citizens with internet access was multiplied by the probability that they became a victim of identity theft, modified by an estimate of the proportion of these crimes being conducted online (estimated at 25 %).

Both methods of calculation provided similar answers, with an average of £1.7 billion per annum. This compares well with the results of other studies by CIFAS, which also made an estimate of £1.7 billion per annum, and the IFSC, which reported a figure of £1.2 billion per annum.

Finally, the costs of scareware and fake anti-virus were calculated from information published by Symantec on the probability of such an attack and its average cost. The resulting figure of £30m was by far the lowest for any type of cybercrime, but it has been identified as an area of growth. The overall estimate for the economic cost of cybercrime to UK citizens is £3.1 billion per annum.

On a different note, the total estimate for industrial espionage is £7.6 billion. The results for different business sectors are shown below:

- aerospace and defence—£1.2 billion per annum—which is due to the large proportion of revenue that companies in this sector derive from large tendering competitions;
- financial services—£2.0 billion per annum—which is due to extremely high transaction volumes and recent share price fluctuations in this sector;

- mining—£1.6 billion per annum—which is due to both the increasing market value of raw minerals and the high level of mergers in this sector at present.

Although the existence of cybercrime in the UK economy appears endemic, efforts to tackle it seem to be more tactical than strategic. It is believed that the potential for reputational damage is inhibiting the reporting of cybercrime. The problem is compounded by the lack of a clear reporting mechanism and the perception that, even if crimes were reported, little can be done. Additional efforts by the Government and businesses to measure and improve their understanding of the level of cybercrime would allow responses to be targeted more effectively.

1.2.3 Middle East

In addition, many international sources are warning that the Middle East is becoming a major source of cybercrime; for example, Saudi Arabia is ranked as the leading country in the region as the target *and* source of malicious activities online; it is also the number one source of malicious attacks in the Gulf Cooperation Council (El–Guindy 2008, p. 17). Egypt is one of the *most phished* countries in the world with about 1,763 phishing incidents, followed closely by other countries in the region such as Saudi Arabia, the UAE and Qatar. It is not hard to see that cybercrimes are increasing in the region due to growth of user base with poor security awareness and the lack of regulations (Ibid). But even normal cybercrime, such as phishing, has its unique characteristics in Middle East. Due to religious motives and political issues in the region, hackers are successfully sending political or religious scams which urge users to open an email attachment, thereby infecting the computer with malware in order to attack Middle Eastern infrastructure targets such as e-commerce websites, banks, telecommunications and government services (Ibid).

Another important factor making the Middle East, and especially the GCC, a source and target of much cybercriminal activity is the growth of international banking and money laundering. The unique opportunities of a quickly developed financial infrastructure allowing anyone to transfer monetary funds to any country, anonymously and through tangled routes caught the attention of cybercriminals (Ibid). Electronic transfer is an efficient tool for concealing sources of money intakes and laundering illegally earned money. There are many well-known online money laundering cases involving victims in the Middle East who have been tricked in order to steal their identity or transfer money from their real accounts using phishing and scams. In one such example, the attacker might send a well-crafted link to users in the Middle East with an email conveying the message that there is a bulk sum of money someone wants to transfer to a UAE bank account. All too often users will reply to this scam, start to interact, and in the end become the victim.

Terrorist motivation plays a dramatic role in cybercrime in the Middle East as a communication tool and a weapon against an enemy (Ibid). Cyber terror is growing in the region due to religious, political and socio-economic issues, such as unemployment. "Jihad Online" claims to use hacking as a technique to carry out jihad against their enemies. Cyber terrorists use their websites in many activities such as psychological warfare, propaganda, recruitment, fund raising, coordination of actions and data mining. One jihad website captures information about users who browse their websites (Ibid). Those who seem most interested in the group's cause or well-suited to carrying out its work are then contacted. Recruiters may also use more interactive internet technology such as online chat rooms and cybercafés, looking for receptive members of the public, particularly young people who have the religious motive that later can be converted to the terrorists' cause. IT professionals who can be influenced to help with the technology may also be sought. Electronic bulletin boards and forums are used as vehicles to reach out to potential recruits.

1.2.4 Asia/Pacific—Japan (APJ)

According to the Symantec Asia—Pacific Internet Security Threat Report, dated April 2010, the US is ranked first for overall malicious activity with 19 % of the total, while China ranked second worldwide, with 8 % (Symantec 2010, p. 3). In (Asia /Pacific—Japan) APJ, China ranked first for malicious activity in 2009, with 32 % of the regional total, down from 41 % in 2008. Due to population size, number of computer users, and high broadband penetration, the United States and China are always likely to rank highly. China has the most broadband subscribers in the world (significantly more than anywhere else in the APJ region) and malicious activity tends to increase in relation to growth in broadband infrastructure (Ibid).

The United States ranked first for originating attacks detected by APJ-based sensors in 2009, accounting for 26 % of all detected attacks, down from 28 % in 2008. Globally, the US also ranked first in 2009 for originating attacks against global targets in 2009, with 23 % of the worldwide total (Symantec 2010, p. 4), and again ranked first for web-based attacks globally in 2009, accounting for 34 % of the worldwide total. China ranked second globally in 2009, with 7 %, which is a decrease from 13 % in 2008. In the APJ region, although China ranked first for web-based attacks in 2009, its 37 % total for this reporting period is a significant decrease from 2008, when it accounted for 79 % of the total for the APJ region (Ibid).

In 2009, South Korea hosted the highest percentage of phishing URLs, with 43 % of the total. This is a substantial increase from 29 % in 2008, when South Korea ranked second behind China, which decreased to 12 % in 2009 from 35 % previously. Of the phishing URLs identified in South Korea in 2009, 91 % targeted the financial services sector (Symantec 2010, p. 5).

In 2009, 21 % of all spam detected worldwide originated in the APJ region. Within the region in 2009, India ranked first for originating spam, with 21 % of the regional total. In 2008, China ranked first, with 22 % of the regional total. Globally in 2009, India accounted for 4 % of spam detected, and ranked third (Ibid).

With regard to bot-infected computers, China ranked second, with 11 % of the worldwide total. In 2009 however, China ranked first for bot-infected computers in the APJ region, accounting for 41 % of the total, which is a double-digit decrease from 58 % in 2008. The decrease in percentage in bot-infected computers in China is partly due to increases elsewhere in the APJ region, specifically in Taiwan and Japan, both of which significantly increased their percentages for bot-infected computers in the region in 2009.

Taiwan had the second highest percentage of bot-infected computers in the APJ region in 2009, with 28 % of the total. This is a significant increase from 2008, when 12 % of the region's bot-infected computers were in Taiwan. Globally in 2009, Taiwan accounted for 7 % of the worldwide total. Taipei, Taiwan was again the top city for bot-infected computers in APJ and worldwide in 2009, with 19 and 5 %, respectively. Taiwan has ranked second in this category in a number of reports (Symantec 2010, p. 10). The high bot activity in Taiwan may be due to the high broadband penetration there. Previously, the Symantec *Global Internet Security Threat Report* attributed this to the increasing levels of fiber-to-the-home/building (FTTH/B) deployment in Taiwan. As noted, malicious activity tends to grow with increased broadband capacity and FTTH/B connections currently provide the highest bandwidth capacities over traditional DSL or cable lines. Japan had the third-highest percentage of bot-infected computers in the APJ region in 2009, with 11 % of the total. This is an increase from 4 % in 2008, when Japan ranked fifth in the region. In 2009, Japan had 3 % of the global total for bot-infected computers, a high ranking that may be explained by its advanced internet infrastructure, as well as by the significant deployment of FTTH/B in the country (Ibid).

Worms were the most common type of malicious code observed in the APJ region in 2009, accounting for 5 % of the volume of the top 50 potential infections. This is an increase from 43 % in 2008, when worms ranked second to Trojans. It is also a higher percentage than the global total for worms of 43 % in 2009 (Symantec 2010, p. 11).

One of the primary contributors to this increase may be the rapid spread of the Downadup (a.k.a., Conficker) worm, which is designed with certain geolocation features that enable it to target specific regions, one of which is China. Eight of the top 10 malicious threats in the region in 2009 were worms, or had a worm component, up from seven in 2008. Moreoever, the volume of worm activity in the region increased by approximately 10 % in 2009. This increased worm activity also explains the degree of percentage decreases in the other threat types in the region in 2009 (Ibid). Most experts believe that common forms of computer related crime are significantly underreported because "victims may not realize that they have been victimized, may not realize that the conduct involved is a crime, or may decide not to complain for reasons of embarrassment or corporate credibility" (Ibid).

Other reasons for the under-reporting of cybercrime is mass victimization (caused by offences such as virus propagation), whereby: "the number of victims are simply too large to identify and count, and because such programs can continue creating new victims long after the offenders have been caught and punished". A final factor complicating the gathering and comparison of national crime statistics is that transnational computer related crimes are, by definition committed in, or have effects in, at least two states, hence risking multiple reporting or no reporting at all. Thus, much of the information we have on cybercrimes is the product of studies and surveys addressed to individuals working in information security. On such a basis, the obvious problem arises that survey results include only the respondents of people who agreed to participate. Before basing critical decisions on survey information, it is important to find out what the response rate was; although there are no absolutes, in general we aim to trust survey results more when the response rate is high. Response rates for telephone surveys however, are often less than 10 %; while response rates for mail and e-mail surveys can be less than 1 %. It is not easy to make any case for random sampling under such circumstances, and all results from such low-response-rate surveys should be viewed as indicating the range of problems or experiences of the respondents rather than as indicators of population statistics.

1.3 Profiling Cybercriminals

People who intentionally abuse and misuse information cover a spectrum of criminals. Although it is impossible to characterize these criminals in a single profile, there are some interesting and useful criminal phenomena that we need to know in order to be effective in protecting information. Several of these characteristics differentiate cybercriminals from white-collar criminals.

1.3.1 Motives of the Cybercriminal

Although there is no way to describe a "typical" cybercriminal, Parker's interviews have revealed a number of common traits among these individuals (Parker 1998, p. 138). They are, for example, typically white-collar criminals who engage in fraudulent business practices associated with their otherwise legitimate occupations, e.g., a broker engaged in the sale of phony stocks. In psychological terms, they can be said to exhibit the differential association syndrome. They also frequently tend to anthropomorphize the computers that they attack, yet they assume that attacking a computer does no harm to people (Ibid). Many cybercriminals exhibit Robin Hood syndrome, rationalizing that they are taking from victims who—in their view—can afford it. This might also be viewed as attempting to achieve a commendable goal, at the expense of harming others.

1.3.2 How Cybercriminals Use the Network

Cybercriminals can use computers and networks as a tool for the crime, or incidentally to the crime itself. Many of the crimes committed by cybercriminals could be committed without using computers and networks. For example, terrorist threats could be made over the telephone or via postal mail; embezzlers could steal company money out of the safe; con artists can come to the door and talk elderly individuals out of their savings in person (Shinder 2010, Online). Even those crimes that seem unique to the computer age usually have counterparts in the pre-internet era. Unauthorized access to a computer is technically different, but not so different in mindset, motives and intent from unauthorized access to a vehicle, home or business office (a.k.a. burglary). Defacing a company's web site is also in many ways very similar to painting graffiti on that company's front door (Ibid). Computer networks have done for criminals the same thing they've done for legitimate computers users: they've made the job easier and more convenient.

Some cybercriminals use the internet to find their victims. This includes scam artists, serial killers and everything in between. Police can often thwart these types of crimes and trap the criminals by setting up sting operations in which they masquerade as the type of victim that will appeal to the criminal. We tend to think of this in relation to crimes such as child pornography and pedophilia, but it's the same basic premise as setting up a honeypot on a network to attract the bad guys (Ibid).

In other cases, criminals use the networks for keeping records related to their crimes (such a drug dealer or prostitute's list of clients) or they use the technology to communicate with potential customers or their own colleagues in crime. Amazingly, a significant number of criminals use their own corporate laptops or email accounts to do this. This is a situation whereby IT professionals may stumble across evidence of a crime inadvertently—including crimes that are not, themselves, related to computers and networks (Ibid).

1.3.3 Types of Cyber Criminals

The cyber criminals consist of various groups and category. This division may be justified on the basis of the object that they have in their mind. The category of cyber criminals are shown below in Table 1.2.

Criminal profiling is the art and science of developing a description of a criminal's characteristics (physical, intellectual, and emotional) based on information collected at the scene of the crime. A criminal profile is a psychological assessment made before the fact—that is, without knowing the identity of the criminal. The profile consists of a set of defined characteristics that are likely to be shared by criminals who commit a particular type of crime. It can be used to narrow the field of suspects or evaluate the likelihood that a particular suspect committed the offence.

Though not quite that easy or certain in real life, criminal profiling is a valuable tool that can give investigations many clues about the person who commits a specific crime or series of crimes. Nonetheless, it's important to understand that a profile—even one constructed by the top profilers in the field—will provide only an idea of the general type of person who committed a crime; a profile will not point to a specific person as the suspect. Although good profiles can be amazingly accurate as to the offender's occupation, educational background, childhood experiences,

Table 1.2 Classification of cyber criminals

S. no	Category	Explanations
1	Children and adolescents between the age group of 6–18 years	The simple reason for this type of delinquent behaviour pattern in children is seen mostly due to the inquisitiveness to know and explore the things. Other cognate reason may be to prove themselves to be outstanding amongst other children in their group. Further, the reasons may be psychological even
2	Organised hackers	These kinds of hackers are mostly organised together to fulfil certain objective. The reason may be to fulfil their political bias, fundamentalism etc.
3	Professional hackers/ crackers	Their work is motivated by the colour of money. These kinds of hackers are mostly employed to hack the site of the rivals and get credible, reliable and valuable information. Further, they are even employed to crack the system of the employer basically as a measure to make it safer by detecting the loopholes
4	Discontented employees	This group include those people who have been either sacked by their employer or are dissatisfied with their employer. To avenge they normally hack the system of their employee

material status, and even general physical appearance, there will always be many individuals who fit a given profile.

In *Scene of the Cybercrime* (Shinder 2002, p. 103), cybercriminals are classified into the following main categories.

1.3.3.1 Hackers, Crackers, and Network Attackers

The network is an important tool that makes white—collar criminals' and scam artists' jobs easier, but it is an absolutely essential tool for hackers. Unless a hacker has physical access to a computer with the Net, it would be impossible for him or her to commit a crime.

Hackers can commit several crimes, such as unauthorized access, theft of data or services, and destructive cybercrimes such as website defacement, release of viruses and DoS, and other attacks that bring down the server or network.

Hackers learn their "craft" in a number of ways: by trial and error, by studying network operating systems and protocols with an eye toward learning their vulnerabilities, and perhaps most significantly, from other hackers. There is an enormous underground network where those new to hacking can get information and learn from more experienced hackers.

There are numerous sites to meet hackers online, and many more that provide tools that can be used for hacking sites. Websites such as the Ethical Hacker

Network, Cult of the Dead Cow, Hacktivismo, Security Hacks, and Darknet provide information and software to discover vulnerabilities and access systems. Of course, almost any network security tool used for testing problems can be used for these purposes. In addition to this, there are newsgroups, mailing lists, online papers and videos that provide guidance and detailed information. Hacker conferences such as DEFCON and the Black Hat Briefings provide real world opportunities for hackers to meet. The hacker culture, furthermore, divides itself into two groups:

- *Black hats* break into systems illegally, for personal gain, notoriety, or other less–than—legitimate purposes.
- *White hats* write and test open source software, work for corporations to help them heighten their security, work for the government to help catch and prosecute black–hat hackers, and otherwise use their hacking skills for noble and legal purposes.

There are also hackers who refer to themselves as gray hats, operating somewhere between the two primary groups. Gray-hat hackers might break the law, but they consider themselves to have a noble purpose in doing so. For example, they might crack systems without authorization and then notify the system owners of the systems' fallibility as a public service, or find security holes in software and then publish them to force the software vendors to create patches or fixes for the problem.

Ethical hacking is a term used to describe hackers who use their skills to hack networks on behalf of the owners. Numerous courses train computer professionals in hacking systems, including the EC—Council's Certified Ethical Hacker (CEH) certification, courses, and even a Master's of Ethical Hacking and Countermeasures degree that is offered by the University of Abertay in Scotland.

If a hacker has the requisite skills, including the social skills necessary to function in the corporate world, ethical hacking can be a lucrative business. Consultants charge companies $10,000 or more to test their security by attempting to hack into their systems and providing recommendations on plugging the security holes that they find.

1.3.3.2 Criminals Who Use the Net Incidentally to the Crime

Some criminals use the network in relation to their crimes, but the Net itself is not an actual tool of the crimes. That is, the network is not used to commit the criminal activity, although it can be used to prepare for or keep records of the criminal activity. Examples of this type of criminality include:

- Criminals who use the Net to find victims
- Criminals who use computers or networks for recordkeeping
- Criminals who use email or chat services to correspond with accomplices.

Even in cases in which the network is not a tool of the crime, it can still provide evidence of criminal intent and clues that help investigators track down the criminals. We discuss each situation in the following paragraphs:

In the first category, criminals go on to use the internet to actually commit the crime, for example, sending electronic chain letters, emailing fictitious notices purporting to be from the victim's ISPs that request their credit card information, or directing victims to a website that tries to sell them products under false pretenses. In these cases, the internet is a tool of the crime, but the initial act of searching out potential victims is not, by itself, criminal. Thus, a pedophile or rapist or other criminals who use the Net to find victims but then commit the criminal activity in the real world are using the internet incidentally to the crime. However, the internet can also be used to step up a sting operation that will turn the tables and lure the criminal into revealing his or her identity to law enforcement.

The second category includes people who engage in non-computer-related criminal activity such as drug dealing, illegal gambling, or other illicit businesses, who use computers to keep financial records, customer lists, and other information related to the criminal activity, whilst simultaneously utilizing the internet to transfer those files to an off-site location where they will be safer from law enforcement.

Transferring business records to a friend's computer or an internet data storage service is not against the law, so internet use is incidental to this criminal activity, even though the files in question might be important evidence of the crime itself.

The last category includes criminals who work in groups—terrorist groups, theft rings, black hat hackers—often use emails and chats in the same way that legitimate users do: to correspond with people they work with. The correspondence itself is not a crime; it is the illegal activity being planned or discussed that is criminal. However, the correspondence can be used not only to show the criminal's intent and help track the offender down, but also, in some cases, to prove the existence of a criminal conspiracy. This is important because if the elements of conspiracy exist, charges can be brought against all members of the conspiracy, not just the person(s) who physically committed the crime.

1.3.3.3 Real-life Noncriminals Who Commit Crimes Online

In some situations, people who are not criminals in real life engage in criminal conduct online. These include accidental cybercriminals and situational cybercriminals. *Accidental cybercriminals* have no criminal intent. They commit illegal acts online because of ignorance of the law or lack of familiarity with the technology. An example is someone who has a cable modem connection or is using the broadband internet access available in some hotels and opens the Network Neighbourhood folder on his computer and sees other computers listed there. Curious, he might click on the icons just to see what happens. If he has stumbled upon a computer on the network that is running a low-security operating system or doesn't require a username and password to log on, and it has network file sharing enabled, he might be able to access the shared files on that computer.

If our hypothetical user is not very technically or legally savvy, he might not even realize that those files are on someone else's private computer. Or he might think that because they're accessible, it is legal to look at them. However, depending on how the state's or country's unauthorized access statutes are written, it might be a crime to access any other computer across a network without permission, even if that computer's users have, perhaps unwittingly, made it technologically easy to do so.

1.4 Challenges for Criminal Justice and Law Enforcement

The internet facilitates the ability of offenders to communicate directly with other likeminded persons as well as future victims through chat rooms, newsgroups, internet relay channels, websites and emails (Fantino 2009, Online). The high volume of offences on the internet and the lack of international boundaries require the cooperation and sharing of information between and among national and international police departments, government legislators, and the public and private sectors (Ibid). This powerful medium is proving to be one of the greatest challenges law enforcement has ever had to deal with. In the following section we shall provide a brief overview of the major challenges in fight against cybercrime.

1.4.1 Transnational Legal Jurisdictions

Domestic legislation is clearly necessary to target cybercrime offenders; however, various problems with the way that cybercrime is now committed make the use of domestic regulation by itself unworkable (Smyth 2007, p. 59). Acts on the internet that are legal in the country where they are initiated, may be illegal in other countries, even though the act is not particularly targeted at that single country (Ibid). Jurisdiction conflicts abound, both negative (no country claims jurisdiction) and positive (several countries claim jurisdiction at the same time). Above all, it is unclear just what constitutes jurisdiction: is it the place of the act, the country of residence of the perpetrator, the location of the effect, or the nationality of the owner of the computer that is under attack? Or, all of these at once? (Ibid).

1.4.2 Evidence Identification and Tracking

The dynamic and distributed nature of cyberspace makes it difficult to find and collect all relevant digital evidence of cybercrimes. Data can be spread over cities, states or even countries. When dealing with the smallest networks, it is feasible to take a snapshot of an entire network at a given instant (Casey 2004). Furthermore, network traffic is transient and must be captured while it is in transit. Once it is captured, only copies remain and the original data are not available for comparison (Ibid). Although the amount of data lost during the collection process can still be documented, the lost evidence cannot be retrieved. Once it is captured, only copies remain and the original data are not available for comparison. Furthermore, open networks contain large amounts of data and sifting through them for useful information can be like looking for a needle in a haystack and can stymie an investigation (Ibid).

1.4.3 Tactics for Evasion

A further set of problems arise where an offender is using encryption. There are several means by which this might take place. In the most common, parts of the suspect's stored data are encrypted—most of the PC is "open" but there are directories, sections, files, or "containers" which hold files, which are encrypted (Sommer 2002). This approach is popular because it is easy to implement and there are relatively large numbers of robust software products available; the computer can be used normally and then specific actions are needed to decrypt the "secret" items (Ibid). Furthermore, some IRC clients support encryption, making it more difficult for investigators to monitor communications and recover digital evidence (Ibid).

Offenders can make it more difficult to locate them on IRC by using the invisibility feature (Casey 2004). This feature does not conceal the offender from other internet users in the same channel, however, so offers only limited protection. One advanced aspect of IRC that some criminals use to conceal their IP address are "bots" (Casey 2004). These programs can work like proxies and are used to perform various tasks from administering a channel to launching denial of service attacks. "Eggdrop" is one of the more commonly used IRC bots and can be configured to use strong encryption that conceals the contents of its logs and configuration files making it necessary to examine network traffic to observe nicknames and passwords (Casey 2004). Finally, cyber offenders who are more technically savvy and are especially interested in concealing their identity, send messages through anonymous or pseudonymous services. When an email is sent through an anonymous remailer, identifying information is removed from the email header before sending the message to its destination (Ibid). The most effective anonymous remailers are quite sophisticated and make it very difficult to determine who sent a particular message. Some remailers keep logs of the actual email addresses of individuals, but many of them will perish than make such concessions, even when illegal activity is involved. There is a possibility that investigators can compel a pseudonymous remailer to disclose the identity of the sender but it requires significant effort since their business is to protect the identity of their users (Ibid).

1.5 The Future of Cybercrime

In the growing world of the internet, both in personal and internet businesses, cybercrime is an ever increasing problem. Punishment for these crimes has become a new field in crime investigation and law enforcement. Cybercrime has taken criminals across borders and limitations that nothing else has been able to, until the advent of the internet. Where a door is left open, the criminal element will find their way in. In this case, the door for crime is the internet.

Cybercrime is vast in scope. It ranges from the individual criminal, to an increasing presence of International Organized Cyber Crime. Scams run rampant across the internet. They fill email boxes and websites, trying to lure unsuspecting victims into their webs of deception. Spy bots and Trojan programs attempt to infiltrate sensitive personal and business information, to gather what they can find, to be used with criminal intent.

Crimes of a more personal nature also abound across the internet. Dating and chat sites are rife with scammers or people playing dangerous games with individual human victims. This social interaction becomes a crime when the person is victimized by cyber stalkers, or people who use the internet to bolster their insecurities, at the expense of acting like a real human being. Regrettably, children are often the victims of such crimes. News headlines, moreover, often include stories of victims to these internet or cybercrimes. Some victims have reacted in personal desperation to violations perpetrated by the online stalkers and game players.

Maintaining security and safety on the internet has become increasingly more complex. Whole companies and businesses exist to deal with the problems.

What's in the future for internet crime and punishment? With every new avenue opening up on the internet comes more possibilities for criminal intent. The difference now, and the future, is that technology and human services are either in place or coming into place, to make these individuals and organizations accountable for their actions. Laws and punishments for even the smallest internet crimes are now on the books, or in the process of being created. Make no mistake, once something is on the internet, it is fact. It is traceable and punishable. No matter how hard someone tries to cover it up, erase it or disassociate from their actions, once the footprint is made, it can't be unmade. Somewhere there is a way to track that footprint. Law enforcement across the globe will enforce it.

The internet has not only drawn people together, it has drawn international crime fighting agencies together in a common purpose. The internet is not a free playground anymore. It is a global arena.

1.6 Summary

It is clear that cybercrime is a growth industry internationally. Precisely because of its international nature, such crimes create many political and jurisdictional problems and problems arising from the incompatibility of criminal and criminal-procedure codes.

Therefore, it is of the greatest importance that Arab countries ratify international documents such as the Convention on Cybercrime. Failing to do this will create "crime shelters" similar to "tax shelters" created by the legislation in certain states.

Furthermore, without the necessary political will and corresponding funding for the required administrative structures, many countries will quickly become an easy target for international cybercrimes.

References

W. Blackstone, *Commentaries on the Laws of England* (University of Chicago Press, Chicago, 1979)

E. Casey, *Digital Evidence and Computer Crime* (Elsevier Academic Press, California, 2004)

R. Crutchfield, *Crime: Readings* (Pine Forge Press, California, 2000)

M. El–Guindy, Cybercrime in the Middle East. ISSA J. 17 (June, 2008)

J. Fantino, Child pornography on the internet: new challenges require new ideas. Police Chief Mag. (2009), http://policechiefmagazine.org. Accessed 23 July 2011

D. Goodman, S. Brenner, The emerging consensus on criminal conduct in cyberspace. Int. J. Law Inf. Technol. **10**(2), 3 (2002)

P. Hitchens, *A Brief History of Crime* (Atlantic Publishing, London, 2003)

D. Parker, *Fighting Computer Crime: For Protecting Information* (Wiley, New York, 1998)

D. Shinder, *Scene of the Cybercrime* (Walthman, Syngress, New York, 2002)

D. Shinder, Profiling & categorizing cybercriminals (2010), http://www.techrepublic.com. Accessed 10 July 2011

R. Smith, P. Grabosky, G. Urbas, *Cyber Criminals on Trial* (Cambridge University Press, Cambridge, 2004)

S. Smyth, Mind the gap: a new model for internet child pornography regulation in Canada (2007), http://www.sfu.ca. Accessed 11 July 2011

P. Sommer, in *Evidence in internet pedophilia cases*. NCS/ACPO Conference, July 2002, Bournemouth, http://www.pmsommer.com. Accessed 01 July 2011

Symantec, Symantec internet security threat report (2010), http://www.symantec.com. Accessed 8 July 2011

UNESCO, *Les Dimensions Internationales du Droit du Cyberespace* (Economica, Paris, 2000)

Part II
Computer System as Target

Chapter 2
Unauthorized Access Offences in Cyberworld

2.1 Emerging Threats: Expected Targets and Forms

Computer hacking is the accessing of a computer system without the express or implied permission of the owner of that computer system (Bainbridge 2004, p. 381). A person who engages in this activity is known as a computer hacker and may simply be motivated by the mere thrill of being able to outwit the security systems contained in a computer. Hackers may gain access remotely, using a computer in his own home or office connected to a telecommunications network (Ibid).

Hacking can be thought of as a form of mental challenge, not unlike solving a cross word puzzle, and the vast majority of hacking activities have been relatively harmless. Sometimes, the hacker has left a message publicizing his feat, reflecting the popular image of a hacker as a young enthusiast who is fascinated by computers and who likes to gain access to secure computer systems to prove his skills to himself or his peers (Ibid).

Computers can be the target of a criminal activity, a storage place for data about a criminal activity and/or the actual tool used to commit a crime (planning criminal activity). One of the most publicized crimes targeting computers involves unleashing a virus through email. A virus is a computer program that disrupts or destroys existing computer systems. A virus spreads rapidly around the world destroying computer files and costing companies and individuals millions in downtime (time when the computers or networks are shutdown). Most viruses are released by hackers as pranks. A hacker is someone who gains unauthorized access to a specific system. Sometimes hackers may target law enforcement or military computers and read or copy sensitive (secret or private) information. Some are concerned that terrorists will unleash viruses to cripple computer systems that control vital transportation networks.

M. Chawki et al., *Cybercrime, Digital Forensics and Jurisdiction*,
Studies in Computational Intelligence 593, DOI 10.1007/978-3-319-15150-2_2

Once the hacker has penetrated a computer system he might do one of several different things. He might read or copy highly confidential information; erase or modify information or programs stored in the computer systems; download programs or data, or he might simply add something, such as a message, boasting of his feat. He might also be tempted to steal money or direct the computer to have goods sent to him.

In days before computers, sensitive information was kept locked away in filing cabinets in locked rooms on the premises of the organization holding the data. This way the sensitive information was safe from being tampered with or copied.

By contrast, information stored on a computer that is linked to a telecommunication system is much more vulnerable. It is analogous to information stored in paper files kept in locked cabinets but left in a public place. It is just a matter of finding the right key to fit the cabinet; not only can a total stranger try the lock, but, often he can spend as long as he likes trying different keys with impunity until he finds one that works.

A recent example of hacking's dangerous effects can be seen in the various botnet conspiracies currently plaguing the country (Department of Justice 2010, Online). As background, "botnets" are "collections of software agents that run automatically" to commandeer massive numbers of computers to allow cybercriminals to conduct large-scale "malicious activity including spreading spam, stealing log-in credentials and personal information or distributing malware to others." (Pinguelo and Muller 2011, p. 132). In one small example, conspirators allegedly created a coded botnet program, which could be used to hack into and control another person's computer. Once transmitted, the program caused the infected computers to log onto a website and wait for commands, allowing the men to control and command the botnet.

With the botnet subject to their every whim, the men accessed, without permission, the user database of T35.net, a website which offered personal and business web-hosting services for thousands of users (Department of Justice 2010, Online). The database contained confidential user identifications and passwords, which the defendants downloaded. Soon thereafter, the men defaced the T35.net website and exposed the customers' user ids and passwords to the public (Ibid).

It is not only small companies that are vulnerable to botnet attacks, as in 2010, large corporations such as Google, Adobe, and several others were victimized by a targeted botnet attack called "Aurora." According to an industry insider, "the Aurora botnet was targeted against large international businesses with the goals of network infiltration, theft of business secrets and modification of critical systems data."

2.2 Criminal Statues

2.2.1 United States

2.2.1.1 The CFAA

The CFAA is a computer security statute aimed at protecting the computers operated by the federal government and banking institutions, and computers linked to the Internet. It creates criminal liability for "trespassing, threats, damage, espionage," and for government computers "being corruptly used as instruments of fraud."

2.2.1.2 Access Device Fraud

Section 1029 outlaws the "production, use, possession, or trafficking of unauthorized or counterfeit access devices." In relation to Cybercrime, the DOJ asserts that the statute could be used to prosecute a cybercriminal who employs "phishing" emails to obtain victims' private passwords and financial account numbers, or where the cybercriminal deals in stolen bank account or credit card information. The penalties for this variety of fraud are severe, including civil forfeiture and prison terms ranging from a maximum of 10 or 15 years for first time offenders, with repeat offenders being subject to a potential 20 years jail sentence.

2.2.1.3 Stored Wire and Electronic Communications and Transactional Records Access

This statute criminalizes the unauthorized access of email and voicemail. The felony version of the crime has five basic elements: (1) intentional access; (2) without or in excess of authorization; (3) access of a facility where an electronic communication service (ECS) was provided; (4) the defendant obtained, altered, or prevented authorized access to a wire or electronic communication while it was in "electronic storage;" and (5) the defendant acted "for purposes of commercial advantage, malicious destruction or damage, or private commercial gain, or in furtherance of any criminal or tortious act…". For first-time offenders who lack the fifth "purpose" element, the maximum penalty is 1 year imprisonment and substantial fines, while repeat violators who lack the "purpose" element, or first-time offenders who commit the act with the "purpose" discussed above, face up to 5 years in prison and heavy fines. Repeat violations that run afoul of the improper purpose element expose the offender to a prison term of up to 10 years, coupled again with extensive fines.

2.2.1.4 Wiretapping and Eavesdropping

In the United States, the use of wire, telephone, or television communication facilities for the purpose of executing a scheme to defraud or obtain money or property by false presences is a federal offence, even where the underlying fraudulent activity is strictly not a federal or state offence.[1] Such use must be proved to be in furtherance of the scheme and not merely incidental to it.

The wiretapping laws predate and hence were not designed to deal with the problem of unauthorized access to a computer, and whether they apply in a given case is inevitably somewhat arbitrary. In *US v Seidlitz* the use of telephone wires to copy computer software led to a wire fraud conviction and in *US v Rifkin* there was a conviction for fraudulent wire transfers. On the other hand in *US v Alston*, where the defendant used keys on a computer to alter personal credit files, thereby enabling unqualified persons to gain cars and other items on credit, the legislation was held not to apply. This is an example of the fortuitous application of existing criminal offences to a new problem, and it cannot be regarded as an appropriate solution to the question of unauthorized access to the computer.

The wiretap statutes current in many other jurisdictions refer expressly to the interception of oral communications or conversations and hence these statutes are inapplicable to the accessing of computers. The Italian Penal Code, for instance refers to communications "between persons", as does the German Penal Code.

In the key Canadian case of McLaughlin, this type of legislation was found to be inadequate and a case of unauthorized access to a computer. In early 1977, Michael McLaughlin made use of a computer owned and operated by the University of Alberta. The university's computer had about three hundred terminals located in and around the campus, connected to the central processing unit by coaxial cable and telephone lines. Mr. McLaughlin used one such terminal to use the computer's central processing unit without authorization.

Michael McLaughlin was charged with theft contrary to s.287(1)(b) of the Criminal Code for "fraudulently and without colour of right" using a "telecommunication facility".[2] Mr. McLaughlin was convicted of theft at trial. The trial judge considered that the central processing unit computer, the memory, the printers and the terminals constituted a telecommunications facility.

The Alberta Court of Appeal held that the accent of the computer facility was on computing and calculation and that the relay or communication aspect was only incidental, and hence allowed the appeal.

The Supreme Court of Canada agreed with the findings of the Alberta Court of Appeal that the university computer facility was not a "telecommunications facility" and that Mr. McLaughlin was therefore not guilty of theft under s.387(1)(b). Mr. Justice Estey, one of the panel of judges at the Supreme Court of Canada, invited Parliament to amend the Criminal Code at page 341 of the reported

---

[1] *US* v *Kelly 507* F Supp 495, 498 n6 (ED Pa 1981), *US v Gallant* 570 F Supp 303 (SDNY1983).

[2] (1979) 19 A.R. 368 (Alta. C.A.); (1980) 2 S.C.R. 331.

decision: "Had Parliament intended to associate penal consequences with the unauthorized use of a computer, it no doubt would have done so in a section of the Criminal Code or other penal statute in which the term which is now so permanently embedded in our language is employed".

2.2.1.5 Pending Federal Legislation

In February 2010, the House of Representatives passed H.R. 4061, the Cybersecurity Enhancement Act of 2010 (Chabrow 2010, Online). The bill, which the Congressional Budget Office (CBO) estimated would cost $639 million from 2010 to 2014 and $320 million thereafter, would have, among other things, assisted the federal government's efforts in developing skilled personnel for its cybersecurity team, organized and prioritized the various aspects of the government's cybersecurity research and development, improved the shifting of cybersecurity technologies to the marketplace, and strengthened the role of the National Institute of Standards and Technology in developing and implementing cybersecurity public awareness and education programs to promote best practices (Ibid).

The Senate's counterpart cybersecurity legislation, S.773: Cybersecurity Act of 2010, was reported on by the Senate Committee on Commerce, Science, and Transportation in March 2010, which recommended that it be considered by the full Senate. The CBO estimates that the Senate bill would cost approximately $1.4 billion from 2011 to 2015. But, although congressional staffers made progress putting together a cybersecurity package that could pass the Senate, and despite the fact that Senate Majority Leader Reid emphasized passing a cybersecurity bill in 2010, industry opposition and partisan bickering stymied the passage of comprehensive reform by the 111th Congress (Bartz 2010, Online). Part of the reason for the delay may be that some members of congress had concerns relating to increased government control of the internet, as the Senate Bill gave the president the power to initiate contingency plans to ensure that vital federal or private services do not go offline in the event of a major cyberattack (Pearlman 2010, Online).

Similar concerns over presidential powers were seen in the opposition to S.3480: Protecting Cyberspace as a National Asset Act of 2010, the so-called "kill switch bill," which would have given the President broad emergency powers to protect critical digital infrastructure following a cyber-attack. Another reason for the failure of comprehensive cybersecurity reform in 2010 was that the White House did little to pressure Congress to move on the bill, in part because the political value of doing so would be minimal considering that cybersecurity receives such little attention from the average voter, and because the administration has been focused on improving the executive branch's readiness (Chabrow 2010, Online).

In 2011 the prospects for comprehensive reform are no better, as the looming presidential election and gridlock caused by a split Congress made one technology expert opine that "the chance of having a comprehensive anything in 2011 with this Congress is slim to none." (Gross 2011, Online).

2.2.2 United Kingdom

Section 1 of the Computer Misuse Act 1990 is aimed directly at hackers who gain access to computer programs or data without any further intention to carry out any further act. It says that a person is guilty of an offence if:

- He causes a computer to perform any function with intent to secure access to any program or data held in any computer, or to enable any such access to be secured;
- The access he intends to secure, or to enable to be secured is unauthorized; and
- He knows at the time when he causes the computer to perform the function that that is the case.

The intent does not have to be directed at any particular program or data, at programs or data of a particular kind, or at programs or data held in any particular computer.

A person guilty of an offence under this section shall be liable:

(a) on summary conviction in England and Wales, to imprisonment for a term not exceeding 12 months, or to a fine not exceeding the statutory maximum, or to both;
(b) on summary conviction in Scotland, to imprisonment for a term not exceeding 6 months, or to a fine not exceeding the statutory maximum, or to both;
(c) on conviction on indictment, to imprisonment for a term not exceeding 2 years, or to a fine, or to both

Section 17 of the Act contains definitions and other aids to interpretation but the Act does not define "computer", "program" or "data". Securing access is widely defined as causing a computer to perform any function, altering or erasing a program or data, copying or moving it to a different location in the storage medium in which it is held, using it or having output from the computer in which it is held. Access to a program includes access to a part of a program.

The offence is committed if the hacker simply intends to make access regardless of whether he/she succeeds but he/she must know, at the time, that the access is unauthorized. Careless or reckless access will not suffice. Because copying is within the meaning of securing access, potentially it can be an offence under section 1 to make a pirate copy of a computer program or other software or to download an unauthorized copy of a computer program.

2.2.2.1 Authorised Access for an Unauthorised Purpose

An employee may have authorisation to use a computer system as a normal part of his duties to his employer. If the employee subsequently uses the system for an auauthorised use—for example, for his own purpose such as carrying out private work or retrieving information for other purposes unconnected with the employment—does the access become unauthorised for the purposes of the Computer Misuse Act 1990? An example of this form of unauthorised use is given by the Audit Commission. A nurse at a hospital had authorisation to use the patient administration system but used it to search for medical details relating to friends and relatives. She then discussed these details with other members of her family. The nurse in question was not prosecuted under the Act, but given a written warning for this breach of patient confidentiality.

Where authorized access is used for an unauthorised purpose, is that access authorised? Two police officers had used the police national computer to gain access to details of motor cars which they wanted for private purposes unconnected with their duties as police officers. They were charged with the unauthorized access to computer material offence under section 1 of the Computer Misuse Act 1990 and convicted at Bow Street Magistrates' Court. However, their appeals to Southwark Crown Court were allowed and then confirmed by the Queen's Bench Divisional Court. In this case, the sole issue was whether the access was authorized. The divisional court held that it was, even though the purpose of the access itself was not authorized.

The court decided that as the police officers were, in fact, entitled to control access to the material within section 17(5) they were authorized to access the computer data even if this was for an unauthorized purpose. As part of their normal duties, the police officers were entitled to access such computer information. But being entitled to access computer material is not the same as being entitled to control access to such material. This is an important and crucial distinction which the court failed to make.

2.2.2.2 Jurisdiction

The international character of some cybercrimes has caused concern about the possibility of criminals escaping prosecution because of jurisdictional issues. In 1985 an accused sent a telex from London intending to divert funds from New York to the accused's account in Geneva. It was held in the Court of Appeal that, had the attempt been successful, the theft would have taken place in New York and the English courts would not have had jurisdiction to try the perpetrator. To prevent this type of problem, the Computer Misuse Act contains complex provisions relating to jurisdiction and extradition in sections 4–9. All that is required is a link with the home country—England and Wales, Scotland or Northern Ireland (as appropriate). That is, the offence must either originate in the home country or be directed to a computer within it: for example, a person from within England attempting to carry out a computer fraud in Sweden, or a person in Italy attempting to hack into a computer located in London.

A final requirement is that of double criminality; that is, if the person operates from within any of the home countries, yet is intending to commit a further offence under section 2 in a different country, that offence is indeed a criminal offence in that other country as well as in the home country. Of course, in most cases this will not present many problems—most countries recognize theft and fraud.

2.3 Other Offences Associated with Hacking

In this section we will some other offences which are associated with hacking.

2.3.1 Unauthorized Modification of Computer Programs or Data

In the United Kingdom, under s3 (1) of the Computer Misuse Act 1990 a person is guilty of an offence if:

(a) he undertakes any act which causes an unauthorised modification of the contents of any computer; and
(b) at the time when he undertakes the act he has the requisite intent and the requisite knowledge.

"Authorization" was applied in a similar manner as it was to the offence of unauthorized access to computer material under s1 of the Computer Misuse Act 1990. Section 17 of the 1990 Act, which dealt with interpretation, gave a thorough, but at the same time broad, definition of modification, chiefly aimed at computer viruses and other forms of software which can cause severe damage to computer systems. In summary, modification is the alteration, erasure or addition of any program or data to the contents of a computer. The addition of a program or data includes the addition of computer viruses and other malicious software. The requisite intent under s3(1)(b) of the Act is defined under s3(2) as an intent to cause a modification of the contents of any computer, and by-so-doing:

(a) impairing the operation of any computer;
(b) preventing or hindering access to any program or data held in any computer; or
(c) impairing the operation of any such program or the reliability of any such data.

The intent need not be directed at any particular computer, program or data, or at programs or data or modifications of a particular kind, or at a particular modification. All that is required is knowledge that the intended modification is unauthorized.

2.3.1.1 Unsolicited Emails

Since adding data to a computer falls within the definition of modification, the question arises as to whether a person who adds data to, for example, a computer disk, without authorization, has the requisite intent. In 2006 the defendant was dismissed from his employment and subsequently sent several hundred thousand emails to his former employer's computer system, which purportedly came from the company's HR manager. Charges were brought under s3 of the Computer Misuse Act 1990 for making an unauthorized modification to the contents of a computer. In the first instance, the judge held that the emails were authorized as the employer's computer system was set up to receive email. It was also held that s3 was designed to prevent the sending of material such as computer viruses. On appeal, the High Court concurred that a person who sets up a computer to receive emails is giving consent to emails being sent to that computer. But the consent does not extend to emails sent not for communication purposes but to disrupt the system.

2.3.1.2 Scope of the Offence

The scope of the s3 offence is fairly broad, but also well-defined, and covers viruses, Trojans, time-bombs (a computer virus which is triggered by a specific date) and logic bombs (a program which will trigger a malicious function if certain conditions are met). A number of successful prosecutions have been brought under s3 for various types of conduct. A freelance typesetter, for example, altered a client's computer with the consequence that the client was denied access, and was consequently convicted under s3.

2.3.1.3 The New s3 Offence as Introduced by the Police and Justice Act 2006

The Police and Justice Act 2006 has introduced a new s3 offence with the new title of "unauthorized acts with intent to impair, or with recklessness as to impairing, operation of computer, etc." The relevant provisions came into force on 1 October 2008. The offence is now committed by a person who undertakes an unauthorized act relating to a computer, knowing that the act is unauthorized, and intending the act to cause, or being reckless as to whether the act will cause, one of the following:

(a) the impairment of the operation of any computer;
(b) the prevention or hindering of access to any program or data held in any computer;
(c) the impairment of the operation of any such program or the reliability of any such data; or
(d) the enablement of any of the things mentioned in paragraphs (a)–(c) above.

As before, the intent or the recklessness need not be directed at any particular computer, program or data, or at programs of a particular kind. The new s3 also has more severe penalties for conviction, the maximum being 10 years imprisonment for a conviction on indictment.

2.3.2 Computer Viruses

A computer virus is a self-replicating program which spreads throughout a computer system, attaching copies of itself to ordinary programs. Often, by the time the virus is detected, many back-up disks also will have been infected. Rumours abound to the effect that viruses are far more likely to be on disks containing pirated software (Bainbridge 2004 p. 399).

There are literally, thousands of viruses and strains of viruses; some are relatively innocuous, like the Italian virus which causes a bouncing ball to appear on screen, but others are more pernicious and may completely corrupt a hard disk. The "AIDS" disk mentioned earlier was distributed as part of a blackmail scheme to over 30,000 organisations world-wide. Other recent viruses causing havoc and considerable expense, estimated to run into billions of dollars world-wide, were the "I Love You", SirCam and Melissa viruses (Ibid).

Obviously, viruses are going to remain a threat in the future but persons responsible for deliberately introducing them into a computer system are clearly guilty of an offence under section 3 of the Computer Misuse Act 1990. This is so even if the perpetrator does not personally carry out the act causing the infection. Section 3 states that the person is guilty if he undertakes any act which causes unauthorised modification. This includes distributing infected disks (Ibid).

Publishing details of how to write computer viruses could fall within the law of incitement; that is, the person publishing the details could be inciting others to commit a section 3 offence. However, there must be an intention on the part of the inciter to bring out the criminal consequences and this may be difficult to prove. In May 1995, an unemployed man who called himself the "Black Baron" became the first person to be convicted of incitement in respect of computer viruses. He was also convicted of 11 charges under the Computer Misuse Act 1990 and the judge warned him to expect a custodial sentence.

There is also a possibility of a charge as an accomplice but, again, intention must be proved. Obvious doubts about the applicability of the law of incitement and accomplices were confirmed by police fears concerning the then imminent publication of a book revealing virus techniques in 1992. The same difficulties apply in regard to access providers on the internet, through individuals responsible for posting details of how to write and spread viruses could be liable to prosecution (Ibid).

Bearing in mind the international nature of the internet, jurisdiction and extradition will be problematic in many cases.

2.4 Summary

Although transnational crime predates the digital age, digital technology has certainly facilitated it, and has posed new challenges for prosecutors. It is safe to assume that cross-border offences are more common today than in years past, and the need for cooperation between criminal justice agencies in different countries is greater than ever before. The traditional mechanisms of international cooperation, including mutual assistance, and other formalities with roots in the 19th century and earlier, are ill-suited to an era where offences can be committed from across the world at the speed of light.

References

D. Bainbridge, *An Introduction to Computer Law* (Longman, New York, 2004)

D. Bartz, Analysis: cybersecurity bill on list for passage this year (2010), http://www.reuters.com/article/idUSTRE6885MF20100909. Accessed 14 July 2011

E. Chabrow, House passes cybersecurity enhancement act (2010), www.govinfosecurity.com/articles.php?art_id=2166. Accessed 14 July 2011

Department of Justice, Another pleads guilty in botnet hacking conspiracy (2010), http://www.cybercrime.gov/smithPlea2.pdf. Accessed 13 July 2011

G. Gross, Congress may be able to tackles tech issues in 2011 (2011), http://www.pcworld.com. Accessed 14 July 2011

A. Pearlman, Federal cybersecurity programs (2010), http://www.fed-soc.org. Accessed 14 July 2011

F. Pinguelo, B. Muller, Virtual crimes, real damages. VJLT **16**(1) (2011)

Chapter 3
Injection of Malicious Code in Application

3.1 Introduction

One of the biggest headaches that come along with networked and internet-connected computers is the absolute requirement of dealing with malicious code attacks (Erbschloe 2004). There is no choice; if your systems are not equipped in some way with antivirus protection, sooner or later some bug will eat them. There is also very little to be gained by whining about how vulnerable computer systems are to malicious code attacks. The unfortunate circumstances that wired societies face can be depicted in the following manner (Ibid):

- Organizations and individuals want computing and communications resources, and want them as cheaply as possible.
- Software and hardware manufacturers work synergistically to meet market demands for cheap but highly functional computing and communications resources.
- The corporate interests that drive cooperation between software and hardware manufacturers have resulted in a marketplace that is dominated by very few companies.
- Market dominance by very few companies has created computing and communications technology ecology with very few species.
- The antithesis to the social forces that drive the dominant companies to cooperate in controlling the marketplace is a counterculture of malicious code writers that revels in embarrassing the corporate giants on their lack of technology prowess.
- The small number of species in the technology ecology makes it easy for the malicious code writers to find vulnerabilities and launch attacks that can spread around the world in a very short time.

Law enforcement agencies and the corporate giants that dominate the computer marketplace label malicious code writers and attackers as criminals and, at times, even as terrorists. The malicious code writers and attackers view the corporate giants

M. Chawki et al., *Cybercrime, Digital Forensics and Jurisdiction*,
Studies in Computational Intelligence 593, DOI 10.1007/978-3-319-15150-2_3

as criminal and parasitic organizations dominated by greedy capitalists. Meanwhile, the governments of the computer-dependent parts of the world are struggling to unify their efforts to fight malicious code attacks, doing so largely under the umbrella of the global war on terrorism.

These circumstances, in the grandest of capitalistic glory, have created a marketplace in which virus protection and computer security product companies have thrived. This labyrinth of social, political, and economic forces have several results, many of which are very embarrassing for modern societies (Ibid):

- Very few malicious code attackers are ever caught by the police.
- Government agencies cannot catch up with malicious code attackers, let alone build a national defense system to stop attacks.
- Large organizations that purchase technology are the prisoners of the dominant technology companies and have little resource or market alternatives.
- Elected public officials, many of whom are the recipients of campaign contributions from the dominant technology companies, are strongly resisting confronting the industry about product liability.

When all is said and done, the burden caused by these collective and converging trends falls on the computer user. State and local law enforcement can do little to help in the computer security and computer crimes realm. The governments, through laws and incident response by federal agencies, are often slow to react to trends. Perhaps most worrisome of all, the dominant technology companies from which you buy products—in designing the products on ever-shorter production and release cycles—do little to protect the end user. If computer users want to keep their computers up and running and keep the malicious code attackers at bay, they need to do two things: (1) take a comprehensive approach to dealing with malicious code attacks, and (2) become a customer of one of the well-established virus protection companies and buy, install, and maintain their products on their computer systems.

3.2 Types of Malicious Code

Malicious code can be categorized into several different types according to the ways in which it infects a system (Griffin 2000):

3.2.1 File Infector Viruses

File infector viruses are those that infect other files or programs on computer systems. They operate in a number of ways. Once the original "host" program is run, the virus can stay resident or "live" inside the system memory (RAM) and infect programs as they are opened, or lay dormant inside another program. Each time that program is run, the virus will infect another program or file.

A second, more complex file infector is one that doesn't alter the program itself, but alters the route a computer takes to open a file. In this way, the virus is executed first, and then the original program is opened. If a program or file that is infected with a file infector virus is passed from one computer to another, over a network or via floppy disk for example, the virus will begin infecting the "clean" computer as soon as the file or program is opened.

3.2.2 Boot Sector Viruses

Whereas file infector viruses infect programs on a computer's hard drive, boot sector viruses can infect hard drives and removable disks, such as floppy disks. The boot sector is an area at the beginning of a hard drive or other disk where information about the drive or disk structure is stored. Symptoms of a boot sector virus may be a computer that is unbootable or gives error messages upon booting. Frustratingly, boot sector viruses may be present with no noticeable problems. One thing should be noted about floppy disks, however. It does not matter whether the floppy disk is a "bootable" disk or not, if the disk is infected with a boot sector virus and the users inadvertently leaves the disk in the drive when he/she reboots the computer, the virus can still be executed.

3.2.3 Macro Viruses

Macro viruses are by far the most common type of malicious code found today. This is due to the popularity of software such as Microsoft Office and others such as Corel Draw, which use macro programming languages extensively in their products. Macro viruses use an application's own macro programming language to distribute themselves. Macro viruses do not infect programs; they infect documents. Macro viruses typically arrive in an infected document, a price list written with MS Word for example. When the file is opened, the virus infects the base template on the victim computer, in this case "Normal.dot". Normal.dot is the "framework" that Word documents are created on. Once this template is infected, every document that is opened from then on will be infected as well, making all documents created or opened in Word a carrier of the macro virus. Macro viruses have been written for most Microsoft Office applications, including Excel, Access, PowerPoint and Word. They can also be found in Lotus AmiPro and Corel products to name a few. One more warning about macro viruses is that they are not platform specific. They can be found and spread through Macintosh, DEC Alpha, Microsoft NT and Microsoft Windows. In other words, just because a computer user received a Microsoft Word file from a colleague using a Macintosh, doesn't mean he/she will not be infected by a macro virus embedded in that document.

3.2.4 Worms

A worm is a piece of code that can make fully functional copies of itself and travel through a computer network and/or across the internet through a number of means. A worm does not attach itself to other programs like traditional viruses, but creates copies of itself, which in turn creates even more copies. The computer "worm" is so-called because of the way in which "rogue" computer code was originally detected. Printouts of computer memory locations would show random "wormhole" patterns, much like that of the patterns on worm eaten wood. The term eventually became shortened and used to describe viruses that could "worm" or propagate across networks and the internet, leaving copies of themselves as they travelled. Worms are prolific due to the fact that most are created using simple scripting languages that can be created with a text editor and become fully functional "programs" under the right conditions. For example, if you were to obtain a copy of the "I Love You" worm and changed the files extension from vbs to txt, you could safely open the file in Notepad and view the structure of the worm. This makes the vbs script worm extremely popular among the script kiddy fraternity, as it takes no (or very little) programming knowledge to modify an existing worm and release it into the wild (when a virus is circulating in the computing community or throughout the internet, it is said to be "in the wild".)

3.2.5 Trojan Horses

Trojan horses are named after the wooden horse from Greek mythology in which Greek soldiers snuck into the city of Troy. Accordingly Trojans are malicious programs that sneak into a victim computer disguised as harmless software. Trojans may also be "wrapped" inside another program so that when the original innocent program is installed, the Trojan program is installed as well. The most commonly described Trojan has a payload that will allow a user on another computer somewhere else in the world to gain full control and access to the files on the victim's computer. In this way, they can be used to launch denial of service attacks such as those that brought down Yahoo! and E-bay early in 2000. Trojan horses typically consist of two parts, the server and the client. The server is the part that is installed on the victim computer. When the server is installed, it allows the remote client to send commands to the computer as if the other person was sitting at the keyboard. The remote attacker can upload and download files, delete and create files on the system, play with the CD drive and generally control most aspects of the victim machine. Most of the approximately 550 known Trojans will send some sort of message to the attacker to let them know the server is running on the computer. Therefore, every time the victim connects to the internet, the person who sent the Trojan will know that the system is online and open for abuse.

3.3 Threats Posed by Viruses, Worms, and Trojan Horses

Whether modern day malicious code takes the form of a virus, worm, or Trojan horse program, it poses a real threat to organizational networks, causing damage that is often debilitating (Schweitzer 2002). The number of people using the internet has increased dramatically in the last few years, and that number is expected to rise as yet more businesses recognize the importance and necessity of a web presence. Computer systems are at risk when a threat takes advantage of a vulnerability and causes harm (Ibid). A threat is any circumstance or event with the potential to cause harm to an organization through the disclosure, modification, or destruction of information, or by the denial of services. The most serious threats at present are the all too familiar viruses and worms that exploit internet-based services such as email or Internet Relay Chat (IRC). These threats have materialized because both individuals and organisations widely adopt and rely on internet-based services in their everyday operations. Understating the different types of threats posed by malicious code is the first step toward defending internet users (Ibid).

3.3.1 Data Corruption and Loss

While an individual computer or server may cost an organization a few thousand dollars, the data stored on that server may be worth the same amount many times over. Every year, the loss of critical and confidential data costs organizations billions of dollars in lost revenue. Nearly any form of code that can be run on a PC can be manipulated toward malevolent action. Viruses, which are designed specifically for the purpose of vandalizing data, are often the cause of file corruption. A virus may inflict damage in different ways (Ibid):

- By destroying existing data: for example, one virus may be designed to destroy only document files while another may be formulated to wipe out the entire contents of a hard drive.
- By manipulating existing data: for example, the WM/Concept macro virus, by interfering with the operation of the "File → Save As" function in Word, forces documents to be saved only as template files.
- By adding unnecessary data: for example, a virus can corrupt a file simply by adding or appending superfluous data to it, rendering the file unusable by the operating system.

Files infected by a virus may become inaccessible and/or exhibit strange or unpredictable behavior. Data loss or corruption affects financial bottom line through missed opportunities and lost revenue. Businesses may find that existing customers or clientele lose faith and no longer do business with an organization that has suffered losses as a result of its failure to safeguard data.

When a company's reputation is tarnished publicly with this information, it may also dissuade new customers from engaging in business transactions with the company. When data becomes corrupted, the options are few. One can either attempt to repair the corrupted data with a utility program designed specifically for that task or restore the data from a trusted backup.

It is far more expensive to replace data than it is to protect it and it is therefore vital that proactive steps are taken to reduce the chance of catastrophic data loss. With the increase in society's reliance on computers, the issue of data security has become even more mission critical. Whether they are working with an important business proposal, patient medical records, or personal tax information, users today are storing more information electronically than ever before. The loss of critical or sensitive data can have both financial and legal consequences for business and home users alike. Modern computer hard drivers can now store enormous amounts of data. Consequently, when viruses or worms succeed in damaging those drivers, the amount of lost data can be staggering.

3.3.2 Data Theft

In today's wired world, confidential information is frequently transmitted and even stored online, sparking privacy concerns and creating new data security challenges. The explosion of internet use has led to the disappearance of conventional boundaries. The internet, while creating new and unique business opportunities, has also introduced new data security threats. *Data theft* is a broad term that encompasses not only the actual theft of information, but also the unlawful handling or viewing of classified data.

One of the tools used in data theft is the *Trojan horse virus*. Hackers use a Trojan horse program to comb PCs or networks for confidential data files. Arriving as an email attachment, this malicious program can compromise a system by opening a "back door" entry into a computer or server. During the summer of 2000, this virus type was demonstrated when besieged AOL employee computers opened the Trojan horse infected email attachment. During the attack, approximately two hundred member accounts were compromised when the Trojan horse program enabled the intruders to access member passwords and credit card information.

It's not surprising that the most widespread and debilitating worms and viruses receive the largest amount of media coverage. Besides their ability to replicate, some computer viruses exhibit another common thread: the ability to deliver a damaging viral payload. Sometimes viruses only display images or messages; other times, they steal or destroy data. However, even if malicious code doesn't contain a destructive payload, it can still become a nuisance by occupying space on the hard drive or memory, resulting in a marked decrease in the overall performance of computers.

When a virus or worm strikes, no matter what its purpose, one aftereffect is lost productivity. The reason is simple: reactive measures require that the virus or worm first be isolated or contained, and the removed from infected computers. However, the cost associated with recuperation from malicious code attacks transcends the simple isolation and removal of malicious code. While virus removal may require numerous man hours and may cost a company hundreds of dollars for just one infected computer, the loss of that computer's productivity can easily cost a company five to ten times that amount. In a company with several hundred (or more) affected computers, productivity loss can have a pronounced and lasting effect on computer profits.

3.3.3 Stuxnet Worm

Stuxnet was originally detected in early 2010 by a computer security company in Belarus, and subsequently found to have infected thousands of industrial control systems worldwide. It was discovered that Stuxnet worm is malware that attacks widely used industrial control systems built by the German firm, Siemens AG (Richardson 2011). The company says the malware was initially distributed via an infected USB thumb drive memory device or devices, exploiting vulnerabilities in the Microsoft Windows operating system. Such systems are used to monitor automated plants—from food and chemical facilities to power generators. Analysts said attackers may have chosen to spread the malicious software via a thumb drive because many SCADA (Supervisory Control and Data Acquisition) systems are not connected to the Internet, but do have USB ports (Ibid). Once the worm infects a system, it quickly sets up communications with a remote server computer that can be used to steal proprietary with a remote server computer, steal proprietary corporate data, or take control of the SCADA system.

Stuxnet contains 70 encrypted code blocks that appear to replace some "foundation routines" that take care of simple yet very common tasks, such as comparing file times and others that are custom code and data blocks. Before some of these blocks are uploaded to the PLC, they are customized depending on the PLC (Falliere 2010).

As of September 25, 2010, Iran had identified "the IP addresses of 30,000 industrial computer systems" that had been infected by Stuxnet, according to Mahmoud Liaii, director of the Information Technology Council of Iran's Industries and Mines Ministry, who argued that the virus "is designed to transfer data about production lines from our industrial plants" to locations outside of Iran. Some parts of Iran's operations ground to a halt, while others survived, according to the reports of international nuclear inspectors. In 2011, it is still not clear that the attacks are over. Some experts who have examined the code believe that it contains the seeds for yet more versions and assaults (Richardson 2011).

According to Symantec Corporation, a maker of computer security software and services based in Silicon Valley, the worm hit primarily inside Iran but also in time appeared in India, Indonesia and other countries. However, unlike most malware, it seemed to be doing little if any harm elsewhere. It did not slow computer networks or wreak havoc. The Symantec study showed that, as of August of 2010, the most affected countries were Iran, with 62,867 infected computers, Indonesia with 13,336, India with 6,552, United States with 2,913, Australia with 2,436, Britain with 1,038, Malaysia with 1,013 and Pakistan with 993 (Ibid).

Stuxnet is very specific about what it does once it finds its target facility. If the number of drivers from the Iranian firm exceeds the number from the Finnish firm, Stuxnet unleashes one sequence of events. If the Finnish drives outnumber the Iranian ones, a different sequence is initiated. Once Stuxnet determines it has infected the targeted system or systems, it begins intercepting commands to the frequency drivers, altering their operation. "Stuxnet changes the output frequency for short periods of time to 1,410 Hz and then to 2 Hz and then to 1,064 Hz", writes Symantec's Eric Chien on the company's blog. "Modification of the output frequency essentially sabotages the automation system from operating properly. Other parameter changes may also cause unexpected effects" (Ibid).

3.3.4 The Number of Threats

A quick survey of competing anti-virus products shows that the number of threats they claim to detect can vary by as much as a factor of two. Curiously, the level of protection each affords is about the same, meaning that more is not necessarily better.

Why? To begin with, there is no industry-wide agreement on what constitutes a "threat". It's not surprising, given that fact alone, that different anti-virus products would have different numbers—they aren't all counting the same thing (Aycock 2010). There are some disputes, for example, as to whether or not automatically generated viruses produced by the same tool should be treated as individual threats, or as only one threat. This came to the fore in 1998, when approximately 15,000 new automatically generated viruses appeared overnight. It is also difficult to amass and correctly maintain a malware collection, and inadvertent duplication or misclassification of malware samples is always a possibility. There is no single clearinghouse for malware (Ibid).

Another consideration is that the reported numbers are only for threats that are known about. Ideally, computers should be protected from both known and unknown threats. It is impossible to know about unknown threats, of course, which means that it is impossible to precisely assess how well-protected your computers are against such threats.

Different anti-virus products may employ different detection techniques, too. Not all methods of detection rely on exhaustive compilations of known threats, and generic detection techniques routinely find both known and unknown threats without knowing the exact nature of what they're detecting (Ibid).

Not all known threats may endanger user's computers. The majority of malware is targeted to some specific combination of computer architecture and operating system, and sometimes even to a particular application. Effectively these act as preconditions for a piece of malware to run; if any of these conditions don't exist—for instance, you use a different operating system—then that malware poses no direct threat to users. It is inert with respect to their computers (Aycock 2010).

Even if it can't run, malware may carry an indirect liability risk if it passes through user's computers from one target to another. For example, one unaffected computer could provide a shared directory; someone else's compromised computer could then deposit malware in that shared directory for later propagation. It is prudent to look for threats to all computers, not just to their own (Ibid).

3.4 Legislative Approaches

In this section, we will see various legislative approaches which are used worldwide.

3.4.1 The American Approach

Before legislation in the eighties, the American courts used common law principles to prosecute computer crime, most often drawing analogies between ordinary crimes and the new situations created by the new technology. It became a difficult task to attempt to analogize virus distribution to traditional common law transgression, such as trespass. The increase in technology use led to further cases and the widespread realization that legislation was required to improve the situation (Klang 2003).

The Computer Fraud & Abuse Act of 1986 replaced the first piece of legislation (The Counterfeit Access Device and Computer Fraud and Abuse Act of 1984) and was a marked improvement in clarity and usability. This new Act specified that "unauthorized access to a government computer" was a felony, and "trespass into a federal government computer" was a misdemeanor. However, several difficulties became clear in its application, for example, the prescription of a too narrow a standard of culpability. The Act required that the virus writer or distributor must "knowingly" or "intentionally" cause the damage, a detail difficult to prove due to the fact that once the virus is released it is almost impossible to know how and where it will strike.

There have also been amendments to the legislation concerning virus regulation in the form of the 2001 Patriot Act. The Patriot Act amends the penalties for hackers that damage computers and also it eliminates mandatory minimum sentences. Prior to the amendment, offenders violating section 1030(a)(5) could receive no more than 5 years imprisonment, while repeat offenders received up to a maximum of 10 years. It was felt that these sentences where inadequate to deal with such offenders, who had created viruses like the Melissa virus which caused huge damage.

Previous law also included mandatory sentencing guidelines with a minimum of 6 months imprisonment for any violation of section 1030(a)(5), as well as for violations of section 1030(a)(4) (accessing a protected computer with the intent to defraud).

The amendment raises the maximum penalty for violations for damaging a protected computer to 10 years for first offenders, and 20 years for repeat offenders. At the same time the amendment removes the mandatory minimum guidelines sentencing for section 1030 violations.

3.4.2 The Council of Europe Convention on Cybercrime

The Convention on Cybercrime includes provisions dealing with illegal access and interception of computerized information of any kind, including data and system interference. Some provisions contained in the draft treaty limit the production, distribution, and possession of the software used by hackers to exploit computer vulnerabilities.

Amongst the many acts which the convention attempts to regulate is the creation and distribution of the computer virus:

Article 4—Data interference

1. Each party shall adopt such legislative and other measures as may be necessary to establish as criminal offences under its domestic law, when committed intentionally, the damaging, deletion, deterioration, alteration or suppression of computer data without right.
2. A party may reserve the right to require that the conduct described in para 1 result in serious harm.

The aim of Article 4 is to provide computer data and computer programs with protection similar to that enjoyed by corporeal objects against intentional infliction of damage. The protected legal interest here is the integrity and the proper functioning or use of stored computer data or computer programs. In para 1, 'damaging' and 'deteriorating' refer to the alteration of computer programs or data. Deletion is equated with the destruction of a corporeal thing since deletion makes data useless or unrecognizable. The concept of suppressing data is the making of data unavailable to the legitimate user. Alteration refers to the modification of existing data and would include the addition of viruses, Trojan horses and logic bombs etc. The actions in Article 4 are only punishable if they are committed without authorization and the offender must have acted with intent.

The second paragraph allows for legislation to include the proviso that criminalization must require serious harm. The concept of serious harm is left up to each legislating state to decide but each state is under obligation to notify the Secretary General of the Council of Europe of their interpretation if use is made of this reservation possibility.

Article 5—System interference
Each party shall adopt such legislative and other measures as may be necessary to establish as criminal offences under its domestic law, when committed intentionally, the serious hindering without right of the functioning of a computer system by inputting, transmitting, damaging, deleting, deteriorating, altering or suppressing computer data.

The purpose of this provision is to criminalize the intentional sabotage which prevents the lawful use of computer systems (here computer systems also include telecommunications facilities, by using or influencing computer data). The attempt is to create a level of protection for the legitimate interests of the users of computer or telecommunications equipment. The term “hindering” refers to any and all actions that interfere with the proper functioning of the system. This could be anything from inputting, transmitting, damaging, deleting, altering or suppressing computer data.

To create criminal sanctions, it is not enough that hindering has taken place; the hindrance must also be of a serious nature. Each state shall be able to define for itself what the level of seriousness may be. The drafters of the convention, however, consider serious “the sending of data to a particular system in such a form, size or frequency that it has a significant detrimental effect on the ability of the owner or operator to use the system, or to communicate with other systems (e.g., by means of programs that generate “denial of service” attacks, malicious codes such as viruses that prevent or substantially slow the operation of the system, or programs that send huge quantities of electronic mail to a recipient in order to block the communications functions of the system)”.

Article 6—Misuse of devices

1. Each party shall adopt such legislative and other measures as may be necessary to establish as criminal offences under its domestic law, when committed intentionally and without right:

   a. the production, sale, procurement for use, import, distribution or otherwise making available of:

      i. a device, including a computer program, designed or adapted primarily for the purpose of committing any of the offences established in accordance with the above Articles 2 through 5;
      ii. a computer password, access code, or similar data by which the whole or any part of a computer system is capable of being accessed,

      with intent that it be used for the purpose of committing any of the offences established in Articles 2 through 5; and
   b. the possession of an item referred to in paras a.i or ii above, with intent that it be used for the purpose of committing any of the offences established in Articles 2 through 5. A party may require by law that a number of such items be possessed before criminal liability attaches.

2. This article shall not be interpreted as imposing criminal liability where the production, sale, procurement for use, import, distribution or otherwise making available or possession referred to in para 1 of this article is not for the purpose of committing an offence established in accordance with Articles 2 through 5 of this Convention, such as for the authorized testing or protection of a computer system.

3. Each party may reserve the right not to apply para 1 of this article, provided that the reservation does not concern the sale, distribution or otherwise making available of the items referred to in para 1 a. ii of this article.

With para 1(a) i the idea was to criminalize the production, sale, procurement for use, import, distribution or otherwise making available of a device, including a computer program, designed or adapted primarily for the purpose of committing any of the offences established in Articles 2–5 of the present Convention. In this section, 'distribution' refers to the active act of forwarding data to others, while 'making available' refers to the act of making available by the placing of said devices online for others to download and use. This also includes the disputable act of linking to a computer virus.

The convention goes quite far in its criminalization of the computer virus. The creation of a virus becomes, under this convention, a criminal offence, the same with the distribution of any virus programs. The interesting issue is that even a hyperlink to a virus entails prosecution for distribution. One cannot help but wonder how far the crime of linking to material can be interpreted as being a criminal act.

3.5 Summary

Today, and for a long time into the future, it is still up to legitimate users to take precautionary measures to ensure the integrity of their systems. The question is when will the law begin to demand a reasonable standard of care from these legitimate users?

Is it fair to cry foul when a virus infects a system and damages data if it was triggered by an employee wishing to read an anonymous love letter or see nude pictures of tennis stars. The effects of the social engineering of the virus must eventually be taken into account if virus legislation is to become well balanced.

By now anyone who opens unknown attachments should know (or should be informed) that they are playing with fire.

The legislation of viruses is a serious affair. The concept itself is shrouded in mystery and fear. This is not a good basis for a balanced and fair debate but tends to be the basis of a witch-hunt. The creation of destructive software must obviously be dealt with swiftly and efficiently by the law in the same manner as any other form of criminal damage. At the same time, new legislation must not be used to give sweeping powers to the courts to remove anything that does not conform to the mainstream of computer usage.

References

J. Aycock, *Computer Viruses and Malware: Advances in Information Security* (Springer, New York, 2010)

M. Erbschloe, *Trojans, Worms, and Spyware: A Computer Security Professional's Guide to Malicious Code* (Butterworth-Heinemann, New York, 2004)

N. Falliere, Stuxnet introduces the first known rootkit for industrial control systems (2010), http://www.symantec.com. Accessed 15 Aug 2011

B. Griffin, An introduction to viruses and malicious code: overview (2000), http://www.symantec.com. Accessed 15 Aug 2011

M. Klang, A critical look at the regulation of computer viruses. Int. J. Law Inf. Technol. **11**(2) (2003)

J. Richardson, Stuxnet as cyberwarfare: applying the law of war to the virtual battlefield (2011), http://www.ssrn.com. Accessed 15 Aug 2011

D. Schweitzer, *Securing the Network from Malicious Code: A Complete Guide to Defending Against Viruses, Worms, and Trojans* (Wiley, New York, 2002)

Part III
Computer System as Tool

Chapter 4
Attempts and Impact of Phishing in Cyberworld

4.1 The Problem of Phishing

Phishing is the act of sending an email to a user falsely claiming to be an established legitimate business, in an attempt to scam the user into surrendering private information that will be used for identity theft. The email directs the user to visit a website where he or she is asked to update personal information, such as passwords and credit card, social security, and bank account numbers that the legitimate organization has already issued. The website, however, is bogus and set up only to steal the user's information. Phishing combines the power of the internet with universal human nature to defraud millions of people out of billions of dollars (Lininger and Dean 2005, p. 1). Nearly every internet user has received a phishing email by now.

As recipients of phishing emails have gradually become wise to the scams, phishing has evolved into 'SMiShing' with offenders sending out computer-generated SMS (cell phone) texts to encourage recipients either to log onto a fake www site, or to call a number purporting to be their bank security department (Wall 2008, p. 20). Even more recently, SMiShing has evolved into 'vishing' which uses VOIP (voice over internet protocol) to send out the messages (Ibid). The main challenge that phishing poses is that the offence is individually minor and tends only to be serious in aggregate, and only then, when the stolen information is used against the owner (Ibid).

On such account, phishing is a serious crime that merits due consideration and adequate prevention and combating. Phishing may be committed in whole or in part by the use of information and communication technologies (ICTs), which dispenses with face-to-face physical contact and allows for distance counters. Historically, fraud involved face-to-face communication since physical contact was primarily the norm (Brenner 2004, p. 6). Even when remote communication—i.e., postal

M. Chawki et al., *Cybercrime, Digital Forensics and Jurisdiction*,
Studies in Computational Intelligence 593, DOI 10.1007/978-3-319-15150-2_4

mail—could be used to set up a fraudulent transaction, it was often still necessary for the parties to meet and consummate the crime with a physical transfer of the tangible property obtained by deceit (Ibid). Nevertheless, the proliferation of ICTs has exerted a profound impact upon the nature and form of the crime, and has altered the mechanisms of crime commission (Chawki and Wahab 2006, p. 4). Nowadays, perpetrators can use fraudulent emails and fake websites to scam thousands of victims located around the globe, and will likely expend less effort in doing so than their predecessors (Ibid). This new form of automated or electronic crime distinguishes online virtual fraud from real-world fraud in at least two important respects (Ibid) (a) it is far more difficult for law enforcement officers to identify and apprehend online fraudsters; and (b) these offenders can commit crimes on a far broader scale than their real-world counterparts.

For a quick summary, phishing attacks trend around 15,000–18,000 for worldwide statistics.[1] Since September 2010, there was a slight decrease although it appears to be 'noise' in the long run. A major change over the past year (since March 2010) is the shift in banks being used. Targets shifted from being primarily regional banks to nationwide banks. About 65 % of phishing attacks target nationwide banks, 30 % target regional banks and only about 5 % go after credit unions. There has been some fluctuation in these numbers, naturally.[2] Each one tends to stay within roughly 5–10 % of the base number, excepted for the noted shift toward national banks. It should also be worth noting that phishing attacks are fairly spread out in the developed world. The United States is the biggest target with 37 % of attacks; the United Kingdom follows with 27 %, then South Africa with 15 %, China with 7 % and Italy with 3 %.[3] Finally, online phishing does carry the seeds of a potential conflict between national legal systems due to the intrinsic transnational and cross-border implications of such crimes, and the relative variation and divergence of national and regional policies dealing with such crimes. Whilst national and international efforts are underway to establish harmonized and consistent national strategies and policies to combat cybercrime, global condemnation as well as adequate universal policies may not be achieved in the near future at least until all states recognize the importance of ICTs and the need for the existence of an adequate regulatory framework (Chawki and Wahab 2006, p. 5).

4.2 Impact and Harm Generated by Phishing

The effects caused by phishing are far-reaching, and include substantial financial losses, brand reputation damage, and identity theft (Trend Micro 2006, p. 5). The independent research and advisory firm Financial Insights estimated that in 2004,

[1] *See* Phishing Statistics, available at (www.brighthub.com), (visited 29/07/2011).

[2] Ibid.

[3] Ibid.

global financial institutions experienced more than $400 million in fraud losses from phishing. U.S. businesses lose an estimated $2 billion a year as their clients become phishing victims. In the United Kingdom, losses from web banking fraud—mostly from phishing—have nearly doubled from £12.2 million in 2004 to £23.2 million in 2005 (Ibid). Nor are financial losses the only impact to businesses. Lost consumer data files and disclosures of unauthorized access to sensitive personal data are taking a toll on consumers' confidence in online commerce. Phishing's effects are also adversely impacting businesses in several other ways, including (Ibid):

- Possible legal implications if employees are attacked on company computers;
- Potential regulatory compliance issues such as HIPAA, Sarbanes-Oxley, and others, if information breaches occur;
- Significant decreases in employee productivity;
- Impacts on IT resources, as phishing emails use storage space and reduce email system performance;
- Impacts on administrators, as IT departments need to patch or repair systems, shut down applications or services, filter Transmission Control Protocol (TCP) ports 139 and 445 at the corporate gateway, apply patches or hotfixes, and instruct corporate users on email protocols.

Gartner also reported that the number of consumers receiving phishing attack emails increased by 28 % in the period from May 2004 to May 2005. Nearly 2.5 million online consumers lost money directly from phishing attacks: of these, approximately 1.2 million consumers lost $929 million in the previous year. In addition, 1 in 20 British users claimed to have lost money to phishing in 2005 (Trend Micro 2006, p. 5). In September 2003, the Identity Resource Center released the conclusions of its survey on the impact of identity theft on its victims. The results were startling:

- Nearly 85 % of all victims learn about their identity theft in a negative manner.
- The average time spent by victims to resolve an identity theft case is approximately 600 h.
- While victims are discovering their identity theft sooner, the elimination of their negative credit report information takes much longer to resolve.

The Identity Resource Center also reported that many victims suffered significant financial and emotional distress from the identity theft. In addition to credit card losses, victims incurred losses that included time lost from work, lost wages from time spent clearing their names, costs related to travel, postage, telephone, obtaining police reports, and numerous other tasks. Some victims never regained their financial health from these identity thefts.

4.3 Mechanisms of Cyberspace Phishing

There are many techniques used by phishers in cyberspace which will be discussed in this section:

4.3.1 Dragnet Method

On January 26, 2004, the Federal Trade Commission filed the first lawsuit against a suspected phisher (Legon 2004, Online). The defendant, a Californian teenager, allegedly created and used a webpage designed to look like the America Online website, so that he could steal credit card numbers (Ibid). In the same year, California federal prosecution prosecuted a 21 years old defendant who used spoof eBay emails and web pages to acquire users' names and passwords, then ran fraudulent auctions on eBay under the victims' names (Rusch 2005, p. 5). This particular crime involved the use of the dragnet method, which comprises the use of spammed emails bearing falsified corporate identification (e.g., trademarks, logos, and corporate names), that are addressed to a large class of people (e.g., customers of a particular financial institution or members of a particular auction site) to websites or pop-up windows with similarly falsified identification. Dragnet phishers don't identify specific prospective victims in advance. Instead, they rely on the false information they include in the e-mail to trigger an immediate response by victims—typically, clicking on links in the body of the email to take them to the websites or pop-up windows where are requested to enter bank or credit-card account data or other personal data.

4.3.2 Rod—and—Reel Method

In a 2004 Connecticut federal prosecution, a young husband and wife team worked together to access chat rooms, use a device to capture the screen names of chat room participants, and send e-mails that directed recipients to disclose their correct billing information, including current credit-card numbers (Rusch 2005, p. 5). Two years later, eight people were arrested by Japanese police on suspicion of phishing fraud by creating bogus Yahoo Japan Web sites, netting themselves 100 million yen ($870 thousand USD). The principals in the scheme then used the credit-card numbers and other personal data to arrange for wire transfers of funds via Western Union, but had others pick up the funds from the money transfer agency itself. In rod and reel method, phishers identify specific prospective victims in advance, and convey false information to them to prompt their disclosure of personal and financial data (Ibid).

4.3.3 Lobsterpot Method

This technique relies solely on the use of spoof websites. It consists in the creation of spoof websites, similar to legitimate corporate ones, that a narrowly defined class of victims is likely to seek out. In lobsterpot phishing, the phishers identify a smaller class of prospective victims in advance, but do not rely on a call to action to redirect prospective victims to another site. It is enough that the victims mistake the spoof website they discover on their own as a legitimate and trustworthy site. In fact, spoof attacks occur at the Protocol layer level. When the spoofer's goal is to either gain access to a secured site or to mask his or her true identity, he or she may hijack an unsuspecting victim's address by falsifying the message's routing information so that it appears to have come from the victim's account instead of his or her own. He or she may do so through the use of "sniffers." Since information intended for a specific computer must pass through any number of other computers while in transit, the data essentially becomes fair game, and sniffers may be used to capture the information en route to its destination. Sniffer software can be programmed to select data intended for any or every computer.

4.3.4 Gillnet Phishing

At West Point in 2004, teacher and National Security Agency expert Aaron Ferguson sent out a message to 500 cadets asking them to click a link to verify grades. The messages appeared to come from a Colonel Robert Melville of West Point. Over 80 % of recipients clicked the link in the message; in response they received a notification that they'd been duped and warning that their behavior could have resulted in downloads of spyware, Trojan horses, and/or other malware. This technique relies far less on social engineering than the preceding techniques. In Gillnet phishing, phishers introduce malicious code into emails and websites. They can, for example misuse browser functionality by injecting hostile content into another site's pop-up window. Merely by opening a particular email, or browsing a particular website, internet users may have a Trojan horse introduced into their systems. In some cases, the malicious code will change settings in user's systems, so that users who want to visit legitimate banking websites will be redirected to a lookalike phishing site. In other cases, the malicious code will record user's keystrokes and passwords when they visit legitimate banking sites, then transmit those data to phishers for later illegal access to users' financial accounts.

4.3.5 Political Phishing

In 2007, Soghoian et al. voiced concerns about the dangers that phishing could pose to online political campaign donations. With tens of millions of dollars raised online each month during the 2008 US presidential election season, the area is a large untapped market for cyber-criminals (Soghoian 2008, p. 5). To demonstrate their point, the researchers purchased two cousin domain names, www.democratic-party.us and www.support-gop.org—misleading domains that include the name of the authentic site to be spoofed. They also cloned the legitimate websites of candidates from both major parties, and used these to create simulation phishing pages (Ibid). The researchers took care to disable the "submission" functionality of each phishing site's donation page. The large submit button on each page was modified to take users to a debriefing page, which would explain the purpose of the website. Any private information (such as a credit card number) entered into the donation form would be discarded as the user clicked on the submit button, without being transmitted to the researchers' webserver. The researchers planned to use the fake phishing sites as a form of phishing education. Visitors would be able to browse around, and even attempt to donate. Since a user's financial information would never be transmitted over the internet by their browser, the risk to the user was minimal (Ibid).

The researchers were never able to put their political phishing sites online as planned, due to legal issues. The university's general counsel, with whom the researchers were working closely, had a number of concerns with the proposal, and based on this advice, the researchers significantly scaled back their project. First, each of the phishing donation pages was modified to include a prominently located bolded disclaimer, which stated:

> This page is a simulated phishing attack. It is not dangerous, but is meant to raise awareness of the dangers of phishing and new deceit tactics. Click here to read more about phishing...

Secondly, the researchers did not put the fake phishing sites online, but had to settle for screenshots of the various donation pages. Instead of being able to freely navigate around a spoof candidate website, visitors could instead skip through a series of slides (Ibid). Finally, the screenshots were uploaded to a third domain, www.politicalphishing.com. This was due to concerns that the general counsel had with the use of the two misleading political domain names. Based on the advice of legal counsel, therefore, the project was significantly scaled back from the researchers' original plan. Instead of being able to create an immersive experience that members of the public could navigate through to learn the risks of political phishing, the researchers had to settle for a research paper and a set of slides (Ibid).

4.4 Using Stolen Financial Information

Offenders use victims' personal information in diverse and numerous ways. Amongst the most common examples of usage are (FTC 2004, Online):

- Calling the credit card issuer to change the billing address on the victim's credit card account, after which the imposter then runs up charges on the victim's account. Because his bills are being sent to a different address, it may be some time before the victim realizes that there is a problem. In the District of Delaware, one defendant was sentenced to 33 months' imprisonment and $160,910.87 in restitution, and another defendant to 41 months' imprisonment and $126,298.79 in restitution for obtaining names and Social Security Numbers (SNNs) of high-ranking military officers from an internet website and using them to apply on-line for credit cards and bank and corporate credit in the officers' names.
- Opening new credit card accounts in the victim's name. When they use the credit cards and do not pay the bills, the delinquent accounts are reported on the victim's credit report.
- Illegal drug or human trafficking. In the District of Oregon, seven defendants have been sentenced to imprisonment for their roles in a heroin/methamphetamine trafficking organization, which included entering the United States illegally from Mexico and obtaining SSNs of other persons. The SSNs were then used to obtain temporary employment and identification documents in order to facilitate the distribution of heroin and methamphetamine. In obtaining employment, the defendants used false alien registration receipt cards, in addition to the fraudulently obtained SSNs, which provided employers with enough documentation to complete employment verification forms. Some of the defendants also used the fraudulently obtained SSNs to obtain earned income credits on tax returns fraudulently filed with the Internal Revenue Service. Some relatives of narcotics traffickers were arrested in possession of false documents and were charged with possessing false alien registration receipt cards and with using the fraudulently obtained SSNs to obtain employment. A total of 27 defendants have been convicted in the case, 15 federally and 12 at the state level.
- Making telephone calls. In *United States v. Bosanac*, no. 99CR3387IEG the defendant was involved in a computer hacking scheme that used home computers for electronic access to several of the largest United States telephone systems and for downloading thousands of calling card numbers (access codes). The defendant, who pleaded guilty to possession of unauthorized access devices and computer fraud, used his personal computer to access a telephone system computer and to download and transfer thousands of access codes relating to company calling card numbers. In taking these codes, the defendant used a computer program he had created to automate the downloading, and instructed his co-conspirators on how to use the program. The defendant admitted that the

loss suffered by the company as a result of his criminal conduct was $955,965. He was sentenced to 18 months' imprisonment and $10,000 in restitution.

- They may counterfeit checks or credit or debit cards, or authorize electronic transfers in another name, and drain the bank account. In one case, a woman got back from her vacation in Las Vegas to find out that her sister has been using her credit card ($584) and had committed fraudulent activities in her name. The victim did not order or receive any of the items that were charged. Furthermore, when she got home, she opened all her mail to discover that she has been denied housing because her sister had stolen her identity.
- Filing for bankruptcy under another name to avoid paying debts they have incurred or to avoid eviction.
- Buying a car by taking out a loan in another name. In United States v. Wahl, No. CR00-285P 138 (W.D. Wash. sentenced Oct. 16, 2000), the defendant obtained the date of birth and Social Security Number of the victim (who shared the defendant's first, last name and middle initial). He then used the victim's identifying information to apply online for credit cards with three companies and to apply online for a $15,000 automobile loan. He actually used the proceeds of the automobile loan to invest in his own business. (The defendant, after pleading guilty to identity theft, was sentenced to 7 months imprisonment and nearly $27,000 in restitution).
- Acquisition of identification, such as a driver's license, issued with their picture, in the victim's name.
- Use of the victim's name to the police during an arrest. In such case if they do not show up at court on the hearing date, a warrant for arrest will be issued in the victim's name.

4.5 Customer Vigilance

Customers may take a number of steps to avoid becoming a victim of a phishing attack that involves inspecting content presented to them and questioning its authenticity. General vigilance includes (Perumal 2008, p. 5):

- If you get an email that warns you, with little or no notice, that an account of yours will be shut down unless you reconfirm billing information, do not reply or click on the link in the email. Instead, contact the company cited in the email using a telephone number or website address you know to be genuine.
- Never respond to HTML email with embedded submission forms. Any information submitted via the email (even if it is legitimate) will be sent in clear text and could be observed.
- Avoid emailing personal and financial information. Before submitting financial information through a website, look for the "lock" icon on the browser's status bar. It signals that your information is secure during transmission.

- Review credit card and bank account statements as soon as you receive them to determine whether there are any unauthorized charges. If your statement is late by more than a couple of days, call your credit card company or bank to confirm your billing address and account balances.

4.6 Summary

Phishing is the act of establishing legitimate business, in an attempt to scam the user into surrendering private information for identity theft. There is various method of phishing including e-mails, broadcast, false redirection etc. There exists various techniques to avoid the phishing but prevention is better than cure. So, we shall avoid such phishing by alerting ourselves. We have also discussed how customer vigilance is equally important.

References

S. Brenner, Cybercrime metrics, old wine, new bottles. Va. J. Law Technol. **9**, (2004)
M. Chawki, M. Wahab, Identity theft in cyberspace: issues & solutions (Lex Electronica), (2006)
Federal Trade Commission, Take charge: fighting back against identity theft (2004), www.ftc.gov. Accessed 01 Aug 2011
J. Legon, Phishing scams reel in your identity (2004), www.articles.cnn.com. Accessed 30 July 2011
R. Lininger, R. Dean, *Phishing, Cutting Identity Theft Line* (Wiley, New York, 2005)
S. Perumal, Impact of cybercrime on virtual banking (2008), www.ssrn.com. Accessed 02 Aug 2011
J. Rusch, The compleat cyber—angler: a guide to phishing. Comput. Fraud Secur. **2005**(1), (2005)
C. Soghoian, Legal risks for phishing researchers (2008), www.ssrn.com. Accessed 01 Aug 2011
Trend Micro, Phishing (2006), www.antiphishing.org. Accessed 29 July 2011
D. Wall, *Hunting, Shooting and Phishing: New Cybercrime Challenges for Cybercanadians in the 21st Century* (The British Library, London, 2008)

Chapter 5
Sexual Harassment in Cyberworld

5.1 Introduction

Sexual harassment is a well-known social problem that affects people at work, school, military installations, and social gatherings (Barak 2005). A worldwide phenomenon (Barak 1997), it has been thoroughly investigated in recent decades in terms of prevalence, correlates, individual and organizational outcomes, and prevention; the range of studies provides an interdisciplinary perspective covering psychological, sociological, medical, legal, and educational aspects of the phenomenon. Although men face harassment, women are the most likely victims.[1]

In many environments on the internet, some users find themselves so captivated by their cyberspace lifestyle that they want to spend more and more time there, sometimes to the neglect of their in-person life (Suler 1999). They may not be entirely sure why they find themselves so engrossed. They can't accurately verbalize an explanation for their "addiction." The humorous substitution of words in the Palace Spa suggests that it is an unnameable "thing"—a compelling, unnameable, hidden force. It's not the chat room or the newsgroup or the e-mail that is eating one's life, but the internal, unconscious dynamic it has ignited (Ibid). Indeed, the internet has two faces, positive and negative (Barak and King 2000). Its positive aspect is that it enables the enrichment and improvement of human functioning in many areas, including health, education, commerce and entertainment. On its negative side, the internet may provide a threatening environment and expose individuals to great risks (Ibid).

In the context of women using the internet, Morahan-Martin (2000) noted the "promise and perils" facing female Net users. Sexual harassment and offence on the internet is considered a major obstacle to the free, legitimate, functional, and joyful

---

[1] *See* Roberts and Mann (2011)

M. Chawki et al., *Cybercrime, Digital Forensics and Jurisdiction*,
Studies in Computational Intelligence 593, DOI 10.1007/978-3-319-15150-2_5

use of the Net, as these acts drive away Net users as well as cause significant emotional harm and actual damage to those who remain users, whether by choice or by duty.

5.2 Harassment in Cyberworld

Sexual harassment is a prevalent phenomenon in face-to-face, social environments (Barak 2005). The harassment of women in the military (Fitzgerld et al. 1999), at work (Richman et al. 1999) and schools are receiving increased attention from both policymakers and the popular media. Sexual harassment is not a local phenomenon, but exists in all countries and cultures, although its perceptions and judgment, and consequently definitions, significantly differ from one culture to another (Barak 2005).

Till (1980) classifies sexual harassment behaviors into five categories: (1) sexist remarks or behavior, (2) solicitation of sexual activity by promise or rewards, (3) inappropriate and offensive, but sanction-free sexual advances, (4) coercion of sexual activity by threat of punishment, and (5) sexual crimes and misdemeanors. Following extensive pilot work, suggestion was made by (Fitzgerald et al. 1995) to change the classification of types of sexual harassment into three different categories: gender harassment, unwanted sexual attention, and sexual coercion.

Gender harassment involves unwelcome verbal and visual comments and remarks that insult individuals because of their gender or that use stimuli known or intended to provoke negative emotions. These include behaviors such as posting pornographic pictures in public or in places where they deliberately insult, telling chauvinistic jokes, and making gender related degrading remarks (Barak 2005).

Unwanted sexual attention covers a huge range of behaviors from being touched without permission, causing fear, or distress, sexual name calling, harassment to rape and sexual assault. Unwanted sexual attention can happen to both women and men and happen between people of the same and opposite sex.

Sexual coercion exists along a continuum, from forcible rape to nonphysical forms of pressure that compel girls and women to engage in sex against their will. The touchstone of coercion is that a woman lacks choice and faces severe physical or social consequences if she resists sexual advances.[2]

All three types of sexual harassment may exist offline or on the internet. However, because of the virtual nature of cyberspace, most expressions of sexual harassment that prevail on the Net appear in the form of gender harassment and unwanted sexual attention. Nevertheless, as sexual coercion is the type that occurs the least often offline, too, it is impossible to conclude whether its relatively low prevalence in cyberspace is a result of the medium or its very nature. In terms of

[2] *See* The Center For Health And Gender Equity (1999)

virtual imposition and assault, sexual coercion does exist nonetheless on the Net, though without, of course, the physical contact.

Gender harassment in cyberspace is very common. It is portrayed in several typical forms that internet users encounter very often, whether communicated in verbal or in graphical formats and through either active or passive manners of online delivery (Barak 2005). Active verbal sexual harassment mainly appears in the form of offensive sexual messages, actively initiated by a harasser toward a victim. These include gender-humiliating comments, sexual remarks, so-called dirty jokes, and the like. This type of gender harassment is usually practiced in chat rooms and forums; however, it may also appear in private online communication channels, such as the commercial distribution through e-mail (a kind of spamming) of pornographic sites, sex-shop accessories, sex-related medical matters (such as drugs such as Viagra and operations similar to penis enlargement).

Mitchell et al. (2003) conducted a national survey among youth aged 10–17, and their caretakers. 25 % of youth had unwanted exposure to sexual pictures on the internet in 2002, challenging the prevalent assumption that the problem is primarily about young people motivated to actively seek out pornography. Most youth had no negative reactions to their unwanted exposure, but one quarter said they were very or extremely upset, suggesting a priority need for more research on and interventions directed toward such negative effects. The use of fileting and blocking software was associated with a modest reduction in unwanted exposure, suggesting that it may help but is far from foolproof. Various forms of parental supervision were not associated with any reduction in exposure (Ibid).

Of the 73 % of respondents who unintentionally entered sex sites, most did so as a result of automatic linking, pop-up windows, and unintended results while using a search engine.

Passive verbal sexual harassment on the other hand, is less intrusive, as it does not refer to one user communicating messages to another. In this category, the harasser does not target harassing messages directly to a particular person or persons but, rather, to potential receivers (Barak 2005). For instance, this type of harassment refers to nicknames and terms attached to a user's online identification or to personal details that are clearly considered offensive. This category also includes explicit sexual messages attached to one's personal details in communication software or on a personal web page (Ibid).

On a different note, some scholars illustrated how flaming creates a hostile environment for women. Although flaming is not necessarily aimed at women, it is considered, in many instances, to be a form of gender harassment because flaming is frequently, typically, and almost exclusively initiated by men. The common result of flaming in online communities is that women depart from that environment or depart the internet in general—what has been termed being 'flamed out'. "Flamed out highlights the fact that the use of male violence to victimize women and children, to control women's behavior, or to exclude women from public spaces entirely, can be extended into the new public spaces of the Internet" (Barak 2005).

A constructive solution has been the design of women-only sanctuaries that offer communities where flaming is rare and obviously not identified with men.

Graphic-based harassment can be active or passive.[3] Active graphic gender harassment refers to the intentional sending of erotic, pornographic, lewd, and lascivious images and digital recordings by a harasser to specific or potential victims. Graphic harassment often occurs via email, instant messaging, redirected/automatic linking, and pop-ups.[4] Passive verbal sexual harassment does not target a specific person, but potential receivers. Nicknames and terms or phrases clearly attached to personal details often encompass this form of sexual harassment (Schenk et al. 2008).

Another area of research that has provided insight into cyber sexual harassment is cyber-stalking. Bocji (2004) defined cyber-stalking as a group of behaviors in which the use of information and communications technology is intended to cause emotional distress to another person. Behaviors associated with cyber stalking include making threats, false accusations (false-victimization), abusing the victim, attacks on data and equipment, attempts to gather information about the victim, impersonating the victim, encouraging others to harass the victim, ordering goods and services on behalf of the victim, arranging to meet the victim, and physical assault (Schenk 2008, p. 83). Many of the same behaviors found in cyber-stalking have been linked to cyber-sexual harassment. Furthermore, behaviors seen with cyber-stalking would be considered sexual coercion on the internet (e.g., explicit threats to harm the internet user or the user's friends or family, threats to harm the user's property, or even following the user's internet activity; Barak 2005).

5.3 Cases and Prevalence of Sexual Harassment in Cyberspace

Many authors refer to sexual harassment on the internet and describe it as prevalent and risky. No empirical survey on the scope and prevalence of sexual harassment on the internet has been conducted to date; accordingly; scholars refer to general impressions and sporadic reports. Cooper et al. (2002) refer to sexual harassment by email as a common abuse of women in workplaces.

Leiblum and Döring argued that the internet provides a convenient vehicle, commonly used, to force sexuality on women through non-social (logging into web pages) and social (interpersonal communication) uses of the Net (Barak 2005). McCormick and Leonard (1996) contended that because of the Net's so-called boys club atmosphere (apparently more relevant up to the mid-1990s than today), this environment is typically characterized by anti-women attitudes and behaviors, including sexual harassment. Following the same conception, Döring states that

[3] *See* Internet Harassment, available at www.unc.edu, (visited 08/08/2011).

[4] Ibid.

men's created sexualized online atmosphere, mainly through pornographic materials, make unwanted sexual advances more likely (Ibid).

Adam (2001) argues that the phenomenon of sexual harassment on the internet highlights the positive process of empowerment that women gain from egalitarian use of the internet. Other scholars (Casey and McGarth) consider cyberspace as an ideal environment for sex offenders to commit sexual harassment and imposition because of its characteristics. Cunneen and Stubbs reported an incident in which an Australian citizen solicited sex among Filipino women through the internet in return for economic privileges. Cooper et al. mentioned the case of an internet user with a paraphilia-related disorder who obsessively used chat rooms to communicate his sexual thoughts to women.

5.4 Enforcement Problems

"Even with the most carefully crafted legislation, enforcing a law in a virtual community creates unique problems never before faced by law enforcement agencies." (Ellison 1998). These problems pertain mainly to international aspects of the internet. It is a medium that can be accessed by anyone throughout the globe with a computer and modem. This means that a potential offender may not be within the jurisdiction where an offence is committed. Anonymous use of the internet, though beneficial in many instances, also promises to create challenges for law enforcement authorities.

5.4.1 The International Stalker

The internet is a global medium regardless of frontiers, and this creates new possibilities for the so-called cyber-stalker. Cheap and easy access to the internet means that distance is no obstacle to the cyber-stalker. A user in France may be stalked by someone on the other side of the world by the click of a mouse. The internet is not a "lawless place" but there are difficulties in applying laws that are made for specific nation states and this would be also true of applying national harassment and stalking laws to the internet.

5.4.2 The Anonymous Stalker

Internet technology creates possibilities for anonymous communications and hence for anonymous cyber-stalking. The identity of a cyber-stalker may, therefore, not ever be revealed or found. The fluidity of identity on the internet has been described as one of its chief attractions. The internet facilitates experimentation with different identities. Users may adopt an on-line persona which bears little, if any, resemblance to his or her real identity.

Pseudonymity is achieved by simply forging or "spoofing" an e-mail header so as to create an online digital persona. For example, Alice can create a new persona for her online participation in Usenet discussion groups with an email address such as Billy–Kid@compuserve.com rather than using her real email address, alice@compuserve.com.

Impersonation of other users may also be possible by faking the header of an email message to make it appear as if it originates from the victim's account. Anonymity on the internet can thus be achieved by using an anonymous re-mailer. Re-mailers are computer services which cloak the identity of users who send messages through them by stripping all identifying information from an email and assigning a random replacement header. The most sophisticated re-mailer technology is called MixMaster which uses public key cryptography, granting unprecedented anonymity to users who wish to communicate in complete privacy. A user who chains together several re-mailers could send communications safe in the knowledge that the trail created would be so complex that it would be impossible to follow. According to Ball, true anonymous re-mailers maintain no database of addresses:

> When messages are re-sent from a truly anonymous re-mailer, the header information is set either to a deliberately misleading address, or to randomly generated characters. There is no record of the connection between the sending address and the destination address. For greater security, many users program messages to pass through five to twenty re-mailers before the message arrives at its final destination. This technique, known as chaining, assures greater security than sending through a single re-mailer. Even if some re-mailers keep secret records of their transactions, a single honest re-mailing system will protect the user. One disadvantage is that unless the sender has identified herself in the body of the message, the recipient has no way to reply to an anonymously sent message.

The ease with which users can send anonymous messages would render legal regulation of online harassment a difficult, if not impossible, task. Tracing a cyber-stalker may prove an insurmountable obstacle to any legal action when the electronic footprints which users leave behind are effectively eliminated by re-mailer technology.

5.5 Legal Regulation

There have been calls in the United States for specific cyber-stalking legislation. It is argued that victims of cyber-stalking are inadequately protected as existing laws are too inflexible to cover online harassment. Since its experiences with regard to the internet tend to be more advanced than those in the UK, this section briefly examines the difficulties experienced in the United States in the legal regulation of e-mail harassment. Such problems are unlikely to be encountered in the UK.

5.5.1 United States

The current US Federal Anti-Cyber-Stalking law is found at 47 USC Section 223. All US states now have legislation designed to deal with real-life stalking, but there have proved to be a number of difficulties in applying these state laws to e-mail harassment.

California was the first state to pass a stalking law in 1990, and all the other states have since followed. The first US State to include on-line communications in its statutes against stalking was Michigan in 1993. Under the Michigan Criminal Code, "harassment" is defined as conduct directed toward a victim that includes repeated or continuing unconsented contact, that would cause a reasonable individual to suffer emotional distress, and that actually causes the victim to suffer emotional distress. Unconsented contact under the Michigan Code specifically includes sending mail or electronic communications to that individual. A number of other US States besides Michigan have anti-stalking laws that include electronic harassment. These states include: Arizona,[5] Alaska,[6] Connecticut,[7] New York,[8] Oklahoma,[9] and Wyoming.[10]

In the US, the constitutionality of state anti-stalking legislation remains undecided. In the past, anti-stalking legislation has been challenged on the grounds of being both too vague and too broad. Michigan was the first state to charge someone with on-line stalking (Ellison 1998). Andrew Archambeau refused to stop sending email messages to a woman he met through a computer dating agency and was charged under Michigan stalking laws in May 1994. Archambeau's lawyers sought to challenge the constitutionality of these anti-stalking laws. In January 1996, Archambeau however pleaded no contest to the stalking charge (Ibid).

McGraw highlights further difficulties in using anti-stalking legislation to combat online harassment. In a number of states, McGraw explains, the language of the statute requires physical activity, thus exempting e-mail harassment. Some state statutes also require a "credible threat" of serious physical injury or death. In such states, email harassment is unlikely to meet this standard. This was true in the Jake Baker case. Using the pseudonym "Jake Baker", Abraham Jacob Alkhabaz, a student at the University of Michigan, posted stories to a newsgroup called "alt.sex. stories". One of Baker's stories described the rape, torture and murder of a woman. Baker used the real name of a fellow student from the University of Michigan for the victim. Baker also corresponded with a reader of the story via e-mail who used a pseudonym of "Arthur Gonda" in Canada. In over 40 emails both men discussed their desire to abduct and physically injure women in their local area. Baker was

---

[5] Arizona Criminal Code (1995): 13–2921.

[6] Alaska Criminal Law Section 11.41.270.

[7] Connecticut Penal Code Section 53a–183.

[8] New York Penal Code § 240.30.

[9] Oklahoma Code (1996): § 21–1173.

[10] Wyoming Code, 6–2–506.

arrested and held without bail and was charged with the interstate transmission of a threat to kidnap or injure another. Though most described Baker as a quiet "computer geek" with no history of violence, the stories he posted on the internet were horrific and disturbing. Nevertheless, a US District Court Judge dismissed the case against Baker, ruling that the threats lacked a specific intent to act or a specific target required under the Michigan stalking law. It was the American Civil Liberties Union's ("ACLU") submission that this was a case of "pure speech".

Most stalking laws require that the perpetrator make a credible threat of violence against the victim; others include threats against the victim's immediate family; and still others require that the alleged stalker's course of conduct constitute an implied threat. While some conduct involving annoying or menacing behavior might fall short of illegal stalking, such behavior may be a prelude to stalking and violence and should be treated seriously.

Online identity stealth blurs the line on infringement of the rights of would-be victims to identify their perpetrators. There is a debate on how internet use can be traced without infringing on protected civil liberties.

In 1985 a University of Michigan student was charged with interstate transmission of a threat, after writing a fictional account on a computer bulletin board of raping and torturing a named classmate. That case was thrown out after a US District Court ruled that the student broke no federal law.

Fourteen years later, security guard Gary Dellapenta pleaded guilty under Californian cyber-stalking law over charges of stalking, computer fraud and solicitation of sexual assault. After his advances were rebuffed by a 28 year old woman he met at church, he engaged in identity fraud by posting ads in her name on AOL and other sites that described his victim's supposed fantasies of being gang-raped. When people responded, he impersonated her in email and in chat rooms, revealing personal information such as her address, her appearance, her phone number and how to bypass her home security system. On at least six occasions, sometimes in the middle of the night, men knocked on the woman's door saying they wanted to rape her.

In 2003 a US woman sought protection after claiming that someone had provided her personal information (including her description and location) to men via an online dating service. The victim discovered the identity theft when she was contacted by a man who said they had arranged a casual encounter through the Lavalife.com dating service. Shortly thereafter she was contacted by a second man following chat with "her" about arranging a separate encounter. She commented "You don't even have to own a computer to be the victim of an internet crime any more".

A year later, a South Carolina man was sentenced to 5 years of probation, 500 h of community service and US$ 12,000 restitution after pleading guilty to offences under federal stalking law. He had admitted sending dozens of email and fax messages to a Seattle city employee who had broken up with him 14 years previously.

5.5.2 United Kingdom

Existing UK laws are sufficiently flexible to encompass online stalking and email harassment. The Telecommunications Act 1984, Section 43, for example, makes it an offence to send by means of a public telecommunications system a message or other matter that is grossly offensive or of an indecent, obscene or menacing character. For the purposes of the Act, a public telecommunication system is any telecommunications system[11] so designated by the Secretary of State and is not confined to British Telecom's telephone system. The Act therefore potentially covers the sending of offensive email messages in some instances.[12] The Act will not apply, however, in cases where the data is transmitted by using a local area network unless part of the transmission is routed through a public telecommunications system.[13] So, whether the Act applies to email harassment will depend upon the telecommunications network used, but the Act is not limited to voice communications.

The Protection from Harassment Act 1997 may also be invoked in cases of online harassment. This Act provides a combination of civil and criminal measures to deal with stalking. It creates two criminal offences, the summary offence of criminal harassment[14] and an indictable offence involving fear of violence.[15] Under Section 2 it is an offence to pursue a course of conduct which amounts to the harassment of another where the accused knew or ought to have known that the course of conduct amounts to harassment. A person commits an offence under Section 4 if he pursues a course of conduct which causes another to fear, on at least two occasions, that violence will be used against him. It is sufficient that the accused ought to have known that his course of conduct would cause the other to so fear on each of those occasions.

---

[11] A "telecommunications system" is defined in Section 4(1) of the Telecommunications Act 1984 as "a system for the conveyance, through the agency of electric, magnetic, electro-magnetic, electro-chemical or electro-mechanical energy, of:

(a) Speech, music and other sounds.

(b) Visual images.

(c) Signals serving for the impartation.....of any matter otherwise than in the form of sounds or visual images...".

[12] The Criminal Justice and Public Order Act 1994, s.92 increased the maximum fine for an offence under section 43 to level 5 from level 3 and made it an imprisonable offence with a maximum term of 6 months. The new sentencing powers brings the penalty more into line with the maximum sentence for transmitting indecent or obscene material through the post (which is 12 months' imprisonment) contrary to Section 11(2) of the Post Office Act 1953.

[13] Also note that the Malicious Communications Act 1988s. 1 creates an offence of sending letters which convey, inter alia, threats with the purpose of causing distress or anxiety. The Act does not however cover telecommunications messages.

[14] A person guilty of this offence is liable to imprisonment for a term not exceeding 6 months: s.2 (2).

[15] A person guilty of this offence is liable to imprisonment for a term not exceeding 5 years: s.4 (4).

The Act also gives courts the power to impose restraining orders on convicted defendants, prohibiting them from further conduct which may be injurious to the victim. Breach of such an order carries a potential sentence of 5 years' imprisonment. Harassment includes both alarm and distress, though harassment, alarm and distress are not specifically defined in the Act and so these terms are to be given their ordinary meaning. The range of behaviour covered by the Act is thus potentially extremely wide. The sending of abusive, threatening emails or the posting of offensive material would constitute an offence under Section 2 of the Act as long as it amounts to a course of conduct (for example, more than one e-mail must be sent) and the offender knew or ought to have known that his conduct amounted to harassment.

According to Home Office Minister Alun Michael, there have been 504 prosecutions under Section 2 of the 1997 Act, which have resulted in 247 convictions. None of these prosecutions were internet related and the majority dealt with neighborhood nuisance issues rather than stalking activity.

5.6 Non Legal Regulation

This chapter has highlighted the limitations of legal regulation of online harassment in cases which involve anonymous and international cyber-stalkers. These limitations in legal regulation are, to some extent, compensated for by the availability of non-legal solutions to online harassment. A number of more suitable ways in which users can both empower and protect themselves from online harassment are discussed below.

5.6.1 Software

New and innovative software programs which enable users to control the information they receive are constantly being developed. There are, for example, technical means by which internet users may block unwanted communications. Tools available include 'kill' files and bozo files which delete incoming email messages from individuals specified by the user. Such tools are included with most available email software packages. There is also specially designed software to filter or block unwanted email messages. These tools, such as CyberSitter and Netnanny, are designed mainly to block the access of children to sexually explicit websites and newsgroups, but can also be used to filter out and block email communications. Some of this software can additionally filter words through the incoming and outgoing email messages. The mandatory use of such software, especially at access level, by libraries and ISPs is criticized within the US because the decisions taken to block certain websites are arbitrary and within the discretion of the private companies that develop these systems. They are also defective, since most of them block such websites as the Middlesex County Club or Mars Explorer, while trying to block the word "sex"; or block websites by looking at the keywords in the meta-tags offered by the individual html files. These tools may be of some use to victims of cyber-stalkers to filter out unwanted messages, nonetheless. In the future, advanced filtering systems which recognize insulting email may also be available.

5.6.2 Education of Internet Users

The education of Internet users is one of the important steps towards self-protection from internet harassment. There are many websites and books which provide information for self-protection from cyber-stalkers for online users. Women are advised, where possible, to adopt either a male or gender neutral user name. Passwords, it is advised, should ideally be a meaningless combination of letters and numbers and changed frequently. Passwords should never be given out and should never be sent out via simple email messages as these are the equivalent of sending traditional "postcards" via postal mail. It is recommended that personal information divulged online be kept to a minimum. Users should regularly check their online profile (finger files) or biography to see what information is available to a potential stalker. To guard against online impersonation, users are also advised to use strong encryption programs such as the Pretty Good Privacy ("PGP") to ensure complete private communications.

5.6.3 Internet Service Providers

Access to the internet is possible through Internet Service Providers ("ISPs"). An individual who receives unwanted email or finds that offensive information about them has been posted on the internet should contact the offender's ISP who may eliminate his or her account. ISPs in Britain, for example, do not provide their customers with anonymous accounts, and every single internet user through the British ISPs or ISPs that provide services within Britain should have identifiable customers. These precautions may assist the police in cases in which they are trying to find the identity of a cyber-stalker who may be accessing the internet and conducting his or her cyber-stalking activities through a British ISP.

5.7 Prevention of Sexual Harassment on the Internet

Generally, three parallel ways of preventing sexual harassment have been advised and executed (Barak 2005): legislation and law enforcement, changing of the organizational–social culture, and education and training of potential victims as well as of potential harassers.

5.7.1 Legislation and Law Enforcement

Legislation seems to be necessary to erect strict, well-defined boundaries for interpersonal sex-related behaviors and to define the sanctions attached to unlawful conduct. Legislation also plays an important social role in communicating the social context of what is accepted and what is not in a given society and, thus, serves as a clear sign of values and morals. Law enforcement is necessary for implementing laws, so that they do not just remain theoretical declarations. Although legislation and law enforcement are of top priority offline and take place in all societies, their usefulness in cyberspace is only partial for a number of well-known reasons. For example, the owner of a computer server, the owner of a website, and different websurfers might be located in different locations, including different countries, and therefore subject to different legal systems. In addition, there is the physical location of the server itself. Thus, a server may physically be located in Aruba and owned by a Brazilian who happens to reside in Morocco; a website accessed by that server offers a chat room hosted by an Egyptian who resides in France; in the chat room, an Australian man sexually harasses, by means of unwanted verbal sexual attention, a Danish female surfer who entered the site. In addition, because of anonymity, high-level privacy, invisibility, and the often lack of individual traces that characterize the internet environment, the efficiency of enforcing the law is at best very partial. Effective means of combating sexual harassment should include two major aspects: changing the culture and norms in which sexual harassment might take place, and educating potential victims and harassers. By focusing on these two independent factors—referring as they do to the situation and to the person components of the sexual harassment equation reviewed above—the behavioral product, it is believed, will be changed.

5.7.2 Changing the Culture and Norms

Attempts at changing the culture in regard to sexual harassment should include the delivery of clear, consistent messages of zero tolerance for sexual harassment and the rejection of any leniency, in addition to stances that are antimisogyny, pro-egalitarian, advocating interpersonal sensitivity and acceptance, respecting minorities, and the like. Educational interventions may include awareness and training workshops for potential victims as well as for potential harassers.

5.7.3 Educating Potential Victims and Harassers

In regard to educating potential victims and harassers, this can be carried out in various forms. For instance, the subject of sexual harassment on the internet can be taught in schools in the framework of programs devoted to smart and safe internet use. Such an educational intervention—offered to children, as well as to any vulnerable population—may review standards of 'netiquette' behavior, together with tips on identifying hostile and malicious communications and impingement of privacy and boundaries. Furthermore, online guides that contain explanations, recommendations, tips, and instructions can be posted on numerous sites to complement previous training and to highlight important issues. It is apparent that educational attempts will not prevent people with high proclivities to sexually harass resulting from their personal needs and dispositions; however, these will perhaps make them aware of possible negative outcomes, to themselves and to victims. It is hoped that for some of these people, educational intervention might change perceptions, attitudes, and values, and at the very least, make them aware of considerations new to them, thereby, contributing to a change in their potential problematic behaviors.

5.8 Summary

The sexual harassment is a big challenge in cyberspace. We can prevent the sexual harassment by legislation and law enforcement, changing of the organizational - social culture and education and training of potential victims as well as of potential harassers. The educational intervention will change perceptions, attitudes, and values, and at the very least, make them aware of considerations new to them, thereby, contributing to a change in their potential problematic behaviors.

References

A. Adam, in *Cyberstalking: Gender and Computer Ethics*, ed. by E. Green, A. Adam, Virtual Gender: Technology, Consumption and Identity (Routledge, London, 2001)

A. Barak, A. King, The two faces of the internet: introduction to the special issue on the internet and sexuality. CyberPsychology Behav. **3**,517–520 (2000)

A. Barak, in *Cross—Cultural Perspectives on Sexual Harassment*, ed. by W. O'Donohue, Sexual Harassment: Theory, Research & Treatment, (Allyn & Bacon, Boston, 1997)

A. Barak, Sexual harassment on the internet. Soc. Sci. Comput. Rev. **23** (1), 77–92 (2005)

P. Bocij, *Cyberstalking: Harassment in the Internet Age and How to Protect Your Family, Westport* (Praeger, CT, 2004)

A. Cooper, G. Golden J. Kent-Ferraro, Online sexual behaviors in the workplace: how can human resource departments and employee assistance programs respond effectively? Sex. Addict. Compuls. **9**, 149–165 (2002)

A. Cooper, I. McLoughlin, P. Reich, J. Kent-Ferraro, in *Virtual Sexuality in the Workplace: A Wake-up Call for Clinicians, Employers and Employees*, ed. by A. Cooper Sex and the Internet: A Guidebook for Clinicians, (Brunner-Routledge, New York, 2002), pp. 109–128

L. Ellison, Y. Akdeniz, *Cyberstalking: The Regulation of Harassment on the Internet*, Criminal Law Review, December, (1998)

L. Fitzgerld, V. Magley, F. Drasgow, C. Waldo, Measuring sexual harassment in the military, sexual experiences questionnaire (SEQ – DoD). Mil. Psychol. **11**, 243–264 (1999)

L. Fitzgerald, M. Gelfand, F. Drasgow, Measuring sexual harassment: theoretical and psychometric advances. B. Appl. Soc. Psychol. **17**,425–427 (1995)

J. Morahan-Martin, Women and the internet: promise and perils. CyberPsychology Behav. **3**,683–692 (2000)

K. Mitchell, D. Finkelhor, J. Wolak, The exposure of youth to unwanted sexual material on the internet: a National Survey of Risk, Impact, and Prevention. Youth Soc. **34**, 330–358 (2003)

J. Richman et al., Sexual harassment and generalized workplace abuse among university employees: prevalence and mental health correlates. Am. J. Public Health **89**, 358–363 (1999)

S. Schenk et al., Cyber—sexual harassment: the development of the cyber—sexual experiences questionnaire. McNair J. **12**(1), (2008)

J. Suler, To get what you need: healthy and pathological internet use. CyberPsychology Behav. **2**, 385–394 (1999)

The Center For Health And Gender Equity, Ending violence against women (1999), www.info.k4health.org . Accessed 08 June 2011

F. Till, *Sexual Harassment: A Report on the Sexual Harassment of Students* (Advisory Council on Women's Educational Programs, Washington DC, 1980)

Part IV
Content-Related Offenses

Chapter 6
Online Obscenity and Child Sexual Abuse

6.1 Introduction

The investigation of cybercrime and the gathering of appropriate evidence for a criminal prosecution, the science of "forensic computing", "digital forensics", or "cyber forensics", can be an extremely difficult and complex issue (Walden 2007, p. 205). This is primarily due to the intangible and often transient nature of data, especially in a networked environment. Technology renders the process of investigation and recording of data for evidence extremely vulnerable to defense claims of errors, technical malfunction, prejudicial interference or fabrication (Ibid). Such claims may lead to a ruling from the court against the admissibility of such evidence.

A lack of adequate training of law enforcement officers, prosecutors and, indeed, the judiciary, will often exacerbate these difficulties. In many countries, substantial efforts have been made over recent years to address this training need, with the establishment of specialized facilities and courses, supplemented by training courses offered by the vendors of forensic applications and services (Ibid). The true problem of the information and communications era therefore seems to be the determination of exactly how much value should be attached to a given piece of information, especially when that information is stored electronically and digitally (Van der Merwe et al. 2008, p. 104).

In the past, when law enforcement investigated a crime, the investigators who analyzed the evidence used to present it to the judge to assist him in taking the correct decision. Criminal investigation training courses always include some forensics in order to understand what prosecutors and judges require in regard to evidence (Wang 2007, p. 217). As the primary focus is on the collection and preservation of effective evidence, at a computer-based crime scene, the highest attention must be given to specifying digital evidence. The major feature which distinguishes cyber-crime from conventional crime is that the evidence at the crime scene is represented in electronic form. This also makes it easier for the criminal to

M. Chawki et al., *Cybercrime, Digital Forensics and Jurisdiction*,
Studies in Computational Intelligence 593, DOI 10.1007/978-3-319-15150-2_6

store, conceal, propagate and remove the information and makes it more difficult to identify him/her (Ibid).

In sexual offences, the internet may be involved in a number of ways, resulting in many sources of digital evidence. It can be an instrumentality when it plays an important role in the commission of the crime, such as enticement of children to engage in sexual activity (Ferraro et al. 2005, p. 4). Durkin also highlights the way in which the internet can be utilized by sex offenders to disseminate images for personal and/or commercial reasons; or to engage in inappropriate sexual communication with children and/or to locate children to abuse. Lanning suggests that abusive images downloaded from the internet may be used to desensitize and/or lower inhibitions in an offender or victim prior to or during an offence (Beech et al. 2008, p. 217). The "stickiness" of data is attributable, in part, to the multiple copies generated by the communications process, particularly in an internet environment, as well as the manner in which data is held and removed on electronic storage media. While the "stickiness" of data will work to the advantage of an investigator, the availability of data may not enable a successful prosecution where the defendant is unaware of its existence. Conversely, the widely held perception that data held on an ICT resource is transient may work to the advantage of a defendant, where he can raise doubt as to the existence or otherwise of relevant forensic data.

6.2 The Role of the Internet in Promoting Child Sexual Abuse

In Britain, the UK Cybercrime Report (2008), commissioned by online criminology firm 1871 Ltd, suggested that more than 3,543,300 offences were committed online in 2007. To date there have been 830,000 instances of sex crimes—where individuals were cyber-stalked or received unwanted sexual approaches by pedophiles (Garlik 2008, p. 5). The arrest of 90 offenders in an online child pornography ring in France highlights the extent of this malaise. The roots of the operation go back to December 2004, when a site containing pornographic pictures and videos of children first came to the attention of French national police. The creator, from the northern town of Clermont (Oise), was just 17-year old at the time when he set up the site. Even though he was arrested in May 2005, the pictures and videos were still on the Net and others were downloading and sharing material from his server; proof as far as the police were concerned, that there was an "organized network in place" for diffusing child pornography. Police seized computers with one of them alone containing more than 30,000 images of young children (France Today 2009).

The impact and role of the internet in the production, dissemination and collection of child pornography have been assessed by a number of authors in recent years and there is a general consensus that "the internet has increased the range, volume, and accessibility of sexually abusive imagery, including child pornography" (Akdeniz 2008, p. 1). The internet has undoubtedly escalated the problem of child pornography by increasing the amount of material available, the efficiency of its distribution, and the ease of its accessibility. The problem of internet child pornography can be divided into three components—the production, distribution, and downloading of images. In some cases, the same people are involved in each stage. However, some producers and/or distributors of child pornography are motivated solely by financial gain and are not themselves sexually attracted to children.

6.2.1 Production of Pornographic Content

This involves the creation of pornographic images. Collectors place a premium on new child pornography material (Wortley 2006, p. 9). However, many images circulating on the internet may be decades old, taken from earlier magazines and films. Images may be produced professionally, and, in these cases, often document the abuse of children in third-world countries (Wortley 2006, p. 9). Abusive images of children available on the internet range from every day or 'accidental' naked images of children to depictions of gross acts of indecency against a child or children, such as penetrative sexual intercourse, sadistic acts of brutality, and bestiality, with victims varying in age from babies to teenagers.[1] Abusive images of children can also be manifested in the form of non-real or pseudo-images, including lifelike virtual abusive images without the use of actual children at all and/or those that mix different aspects and/or combinations of separate pictures to suit the user's preferences (Beech et al. 2008, p. 221). Four typical methods are used in the creation of pseudo-images: (1) an image of a child is inappropriately sexualized (e.g., clothes removed); (2) aspects of a sexualized image of an adult is given child-like qualities (e.g., reduction in breast size, removal of pubic hair); (3) an image of a child is superimposed onto a sexualized picture of an adult or child (e.g., a child holding a toy can be superimposed in a way that makes it appear that the child is holding a man's penis); or (4) a montage of abusive images can be created (Beech et al. 2008, p. 221).

6.2.2 Distribution and Sharing

The World Wide Web is a very attractive way of distributing information. It is also possible to set up a website that can generate revenue by selling services by subscription. The distribution of child abuse material are facilitated through

[1] Researchers found that exposure to images of inappropriate sexual conduct with minors, not to mention actual sexual abuse of children, is damaging to the young child's development. *See* Preston (2009), All Knowledge is not Equal: Facilitating Children's Access to Knowledge by Making the Internet Safer, *International Journal of Communications Law & Policy*, issue 13, p. 118.

commercial websites, user- generated websites and peer-to-peer/file sharing network (Kierkegaard 2008, p. 42). Offenders buy sexual pictures of minors and one can even order a live online molestation of a real child and infant for viewing. The prevalence of home video production facilitates the ease of making and posting sexual images online. Internet sex trading, where teenagers are offered goods or money in exchange for sexual favors, is on the rise (Kierkegaard 2008, p. 42). The international policing agency Interpol's Child Abuse Image Database (ICAID)—a global database for the forensic analysis of digital images of child abuse—currently contains more than 520,000 images and has been used to identify 680 victims worldwide (Elliott et al. 2009, p. 181).

In United States v. Reedy 2000, US Postal Inspectors found the Landslide website advertising child pornography photos. The Texas company associated with the site, Landslide Productions, Inc., was owned and operated by Thomas and Janice Reedy. The US Department of Justice estimated that the Reedys made more than $1.4 million from subscription sales of child pornography in the 1 month that the Landslide operation was in business. Customers could subscribe to child pornography websites through a Ft. Worth post office box, or via the internet (Casey 2004, p. 483). Landslide also offered a classified ads section on its site, allowing internet users to respond to personal ads for child pornography. Although related digital evidence was located in Russia and Indonesia, when investigators obtained Thomas Reedy's computer they found more than 70 images of child pornography and a list containing the identities of thousands of Landslide customers around the world. Thomas Reedy was sentenced to life in prison, and Janice Reedy to 14 years in prison (Casey 2004, p. 484).

On the other hand, a new market for sex work has developed online with the advent of live sex shows broadcasted via webcams (Doring 2009, p. 1094). Some professional female sex workers have reported that their activity in cybersex shows is much more comfortable and safe than the prostitution they previously practiced on the street or in hotels. A potential risk is faced by minors who voluntarily chose to enter into the seemingly unproblematic online sex business with excessive haste, overestimating the financial rewards, while underestimating the negative psychological and social effects. To overcome police tracing, some websites use redirection to forward the customer to a completely different server so that law enforcement must retain alert and verify which sever they are connected to when examining digital evidence.

Newsgroups also provide internet users with a forum to discuss their sexual interests in children and to post child pornography. This service consists of several tens of thousands of themed "groups" in which any of a very large number of topics of interest are discussed offline. Participants do not interact in real time but "post" messages of interest to the group, which others may comment upon. Participants pop into the service every now and then to see how the discussions are progressing. Offenders may use this global forum to communicate with a huge audience, to exchange information and to commit crimes, including defamation, harassment, stalking, and solicitation of minors (Casey 2004, p. 485). A system of news-servers and a particular internet protocol takes care of worldwide distribution. A small number of newsgroups are devoted to pedophilia.

On October 16th, 1996, Sharon Rina Lopatka, an internet entrepreneur in Hampstead, was killed in a case of apparent consensual homicide. Lopatka was tortured and strangled to death by a man who she met on the internet first through Usenet and then in a BDSM channel on IRC. Interestingly, nobody who knew her in person, including her husband, suspected that she was involved in this type of activity. On January 27th, 2000, the offender pleaded guilty to voluntary manslaughter, as well as six counts of second-degree sexual exploitation of a minor. He was sentenced to 36–53 months in prison and 21–26 months for possession of child pornography.

In addition P2P file sharing technology provides internet users with the potential to share local files with a potentially unlimited number of other users. Although one of the main drivers for the development of the technology was the sharing of music (MP3) files, as in Napster, offenders have also found the technology convenient in their own activities. When a file is being downloaded from a peer, the associated IP address can be viewed using netstat. However, some peer-to-peer clients can be adjusted to connect through a SOCK proxy to conceal the peer's actual IP address. KazaA has one feature that can be beneficial for law enforcement. Whenever possible, it obtains files from peers in the same geographic region. Therefore, if investigators find a system with illegal materials, there is a good chance that it is nearby (Casey 2004, p. 489).

Furthermore, live conversations between users on the internet exist in many formats and take place 24 h a day. One of the largest chat networks is Internet Relay Chat (IRC), started in 1988 (Casey 2004, p. 486). IRC can be accessed by anyone on the internet using free or low-cost software. On "chat", participants have a significant degree of anonymity, using nicknames and often adopting online personalities different from their real ones. Again, for pedophiles an additional attraction is the sense of community. Exchange of files is normally achieved by "going DCC", that is, leaving the IRC server system and setting up a Direct Computer to Computer link—IRC client software usually allows users to do this by means of a simple click of the mouse.

Aside from chat networks offences are facilitated through online social networking sites. *MySpace* and social networking sites like it offer thriving communities where young people engage in countless hours of photo sharing. In addition to *MySpace*, other social networking and blogging sites such as *Friendster.com*, *Facebook.com* and *MyYearbook.com* allow users to post pictures, videos, and blogs, as well as supporting email and instant messaging (Kierkegaard 2008, p. 43). However, the structure of such sites does vary. *MySpace* is open to anyone, and has loose age restrictions. *Facebook* users are encouraged and often required to register using their real name. On MySpace, people talk by creating profiles: a page on the service's website which can feature a picture, blurb about oneself, a web log (or 'blog'; basically, an online diary), and other information. The free service also features instant messaging. Users can create their profiles and ask others to exchange materials (Kierkegaard 2008, p. 43).

Finally, online file storage and transfer technologies may be used in sexual activities such as inappropriate file exchange with minors and in trafficking child pornography and pedophiliac text stories. File transfer is performed using a variety of protocols some of which are Simple File Transfer Protocol (SFTP), File Transfer Protocol (FTP), Data Transfer Protocol (DTP), and even HTTP (Penna et al. 2005, p. 11). There are three common methods for storing data online: on the users ISP server; with online file storage providers; and using a user created and provided online storage server. Anonymizing techniques may be used to provide ISPs and online storage providers with false information (Penna et al. 2005, p. 12).

6.2.3 Consumption of Pornographic Content

This involves accessing child pornography via the internet. The images do not need to be saved to the computer's hard drive or to a removable disk to constitute downloading (Wortley 2006, p. 12). In some cases a person may receive spam advertising child pornography, a pop-up link may appear in unrelated websites, or he may inadvertently go to a child pornography website (e.g., by mistyping a key word). In most cases, however, users must actively seek out pornographic websites or subscribe to a group dedicated to child pornography (Wortley 2006, p. 12). In fact, it has been argued that genuine child pornography is relatively rare in open areas of the internet, and, increasingly, those seeking to find images need good computer skills and inside knowledge of where to look. Most child pornography is downloaded via newsgroups and chat rooms. Access to websites and online pedophile groups may be closed and require paying a fee or using a password (Wortley 2006, p. 12).

6.3 Criminal Offenses Against Children

Online content and internet activities with sexual character are widespread. The current section considers internet sexuality and its effects on children. In this regard, *four central areas of online sexuality* may be distinguished. These have already been established outside of the internet and have been traditionally committed in physical space. The internet offers new configurations and possibilities of engaging children in these different areas of behavior:

6.3.1 Cybering

The term "cybering" has become a catch-all to describe a variety of computer based sex-related behaviors (Delmonico et al. 2001, p. 4). When engaging in cybering, offenders seek to stimulate children sexually by exchanging explicit digital texts, images, and/or video—often while masturbating (Doring 2009, p. 1095). Victims can be found in various online chat rooms, online communities, online games, or virtual worlds (e.g., Second Life). Cybering provides offenders with the opportunity to collect new sexual experiences and engage in sexual activities with a diverse range of children in a relatively safe and playful setting, behaviors contributing to sexual empowerment (Doring 2009, p. 1095).

There are relatively few large-scale quantitative studies concerning the prevalence of cybering and even fewer national U.S.-based studies.[2] According to "Youth Internet Safety Survey" Surveys (YISS) 13–19 % of youth have experienced some form of cybering in 2007.[3] Given the anonymity of communication, it is often difficult for youth to assess the age of solicitors, but youth reported that they believed that 43 % of solicitors were under 18.30 % were between 18 and 25.9 % were over 25, and 18 % were completely unknown.[4]

Studies demonstrate that violence is rare in internet-initiated sex crimes. The evidence from the N-JOV Study (Wolak et al. 2008, p. 119) suggests that online child molesters are not among that minority of child offenders who abduct or assault victims because they have sadistic tendencies or lack the interpersonal skills to gain the confidence and acquiescence of victims (Wolak et al. 2008, p. 119). Most online child molesters are patient enough to develop relationships with victims and savvy enough to move those relationships offline (Wolak et al. 2008, p. 119). They know what to say to teens to gain their trust, arouse their sexual interest, and maintain relationships through face-to face meetings (Wolak et al. 2008, p. 119). Abduction is also rare. None of the victims in the N-JOV Study were abducted in the sense of being forced to accompany offenders (Wolak et al. 2008, p. 119). However, about one quarter of the cases started with missing persons reports because victims ran away to be with offenders or lied to parents about their whereabouts. So, in many cases, abduction may have been feared (Wolak et al. 2008, p. 119). On August, 4th 2009 an offender from Wayne Country, Michigan was arrested on charges of cybering.[5] The suspect logged onto his laptop, exposed himself on web-cam and performed a sex act over someone he though was a 13 year old girl, but who turned out to be an undercover cyber-cop instead.[6]

[2] *See* Schrock and Boyd, *Online Threats to Youth: Solicitation, Harassment, and Problematic Content*, Literature Review by the Research Advisory Board of the Internet Safety Technical Task Force, Berkman Center for Internet & Society, available at: <www.zephoria.org>, [visited 01/08/ 2011].

[3] Ibid.

[4] Ibid.

[5] Available at <www.idletechie.com>, [retrieved 01 / 08/ 2011].

[6] Ibid.

6.3.2 Online Grooming

Cyber grooming is a course of conduct enacted by a suspect pedophile which would give a reasonable person cause for concern that any meeting with a child arising from the conduct would be for unlawful purposes (Marsh and Gaynor 2009, p. 161). In the process of grooming, the perpetrator creates conditions which will allow him to abuse children while remaining undetected by others, and the child is prepared gradually for the time when the offender first engages in sexual molestation. Offenders may groom children through a variety of means. For example, an offender may take a particular interest in the child and make him or her feel special. He may well treat the child emotionally like an adult friend, sharing intimate details about his sex life and adult relationship. Another grooming technique is through the gradual sexualization of the relationship (Quayle 2008, p. 21). Offenders thus test the child's reaction to sex by bringing up sexual matters or having sexual materials around and sexualize talking.[7] In essence, grooming involves meeting a child with the intent to commit a sexual offence; but no offence might ever in fact be committed.[8] It is conceivable that an attempted offence might be charged even where no actual meeting had taken place so the *actus reus* would simply be online conversations.[9] In the offender's worst case scenario, the conversations might not even be with areal child but with a police investigator posing as a child.[10] The advantages of enabling police intervention as early as possible to protect the children involved are obvious, but, again, where should the line be drawn?[11] Recently, Southwark Crown Court sentenced Sarah Wilson (21) for grooming a child for sex. The court was told that Wilson, targeted the victim in an internet chat room and quickly began playing on the youngster's vulnerability over a six-month period, then quickly escalated into a multitude of texts and phone calls becoming more and more sexually graphic.

[7] *See* Parkinson, *Family Law and Parent—Child Contact : Assessing the Risk of Sexual Abuse* , MULR 15, available at: <http://www.austlii.edu.au/au/journals/MULR/1999/15.html>, (retrieved 02 August 2011).

[8] Available at <www.idletechie.com>, [retrieved 02 August 2011].

[9] Ibid.

[10] Ibid.

[11] Ibid.

6.3.3 Age Play

Even 'Second life' is not immune from sexual offences. In everyday language, 'Second Life' is often referred to as an online computer game.[12] 'Avatars'[13] are frequently called 'players' and the conditions set up by Linden Lab[14] are considered the 'rules of the game' (Hoeren 2009, p. 3). The established Second Life practice of so-called "age play", in which users request sex with other players who dress up as child avatars has encouraged a growth in players posing as children in order to make money (Kierkegaard 2008, p. 44). Age play is an in-world sexual activity between a child avatar and an adult avatar. Sex is an important feature in Second Life. Participants can make their avatars look like anything they want, and create software renderings of whatever equipment they want to use (Kierkegaard 2008, p. 44). They even go to the extent of actually purchasing scripts and making the avatars engage in simulated sex. In 2009, Linden Lab created a new rating of "Adult" to encompass more extreme violent and sexual content (Second Life Website). According to these plans, any businesses located on the mainland [15] will be moved to a new collection of 'Sims' created for the purpose. Those on privately owned Sims will be required to accept the Adult rating [16] or moderate their content. Residents who are not age-verified (as over-18) will not be able to access such content. The Adult rating will only apply to content or services which are advertised or publicly promoted (Second Life Website).

6.3.4 Exposure to Online Obscenity

Cyber pornography/obscenity, as the term suggests, is the publication or trading of sexually expressive materials within cyberspace (Wall 2001, p. 6). The cyber porn/obscenity debate is very complex because pornography is not necessarily illegal.

---

[12] *See* < http://secondlife.com>.

[13] An avatar is a computer user's representation of himself/herself or alter ego, whether in the form of a three—dimensional model used in computer games, a two—dimensional icon used on internet forums and other communities. It is an "object" representing the embodiment of the user. The term can also refer to personality connected with the screen name, or handle, of an internet user. *See* <www.wikipedia.org >.

[14] Founder of Second Life.

[15] Premium membership allows the resident to own land, with the first 512 m² (of Main Land owned by a holder of a Premium account) free of the usual monthly Land Use Fee. The users pay US $ for their first 65,536 m^2. Any land must be purchased from either Liden Lab or a private seller.

[16] Linden Lab has implemented a new account verification system. Users that want to access adult regions and search results will have to authenticate their accounts by having payment information on file or by using Linden Lab's age verification system. Available at <www.pcworld.com>, (retrieved 02 August 2011).

(Wall 2001, p. 6). Child pornography may not directly physically harm youth each time it is viewed by an adult.[17] However, youth are harmed in the creation of images and video of illegal sexual acts, and child pornography perpetuates the idea that sexual relations with children by adults are acceptable. Those who view child pornography, for instance, may erroneously believe that the children involved are voluntary participants who enjoy the act, failing to recognize a power differential.

In Europe, individuals consume images daily through the various facets of the mass media that might be called as obscene in Islamic countries like Egypt, for example. The total amount of abusive images of children available on the internet at any one time is difficult to quantify due to the inherently dynamic nature of the system and the covert nature of the material (Beech et al. 2008, p. 218). Around 42 % of U.S. youth aged 10–17 encountered sexualized content online in 2006, a significant increase from previous years.[18] The material was witnessed through either wanted exposure, unwanted exposure, or both. Exact statistics on how pervasive pornographic content is on the internet has been much-disputed but does not appear to be as pervasive as initially thought.[19] Unwanted exposure comes from "spam" emails, the mis-typing of URLs into a web browser, and key word searches that "produce unexpected results".[20] In YISS-2, 34 % of youth reported either only wanted exposure or both unwanted and wanted exposure. Wanted exposure is also indicated by the 19–21 % of minors who deliberately visited a pornographic website.[21] Rates of exposure vary according to each country, and in some cases were reported to be higher than in the U.S. In addition, increased overseas rates could be due to increased acceptance of sexualized topics, fewer technical measures such as blocking sites, and varying cultural and home environments. For instance, in a survey of 745 Dutch teens aged 13–18, 71 % of males and 40 % of females reported exposure to adult material in 2006.[22]

Child molesters use both adult pornography and child pornography in the grooming process, albeit for different purposes. Adult pornography is most often used to arouse the victim and break down the child's barriers to sexual behavior. Child pornography is also used to break down the child's barriers to sexual behavior, but serves the additional purpose of communicating the child molester's sexual fantasies to the child. Repeated exposure to both adult and child pornography is intended to diminish the child's inhibitions and give the impression that sex between adults and children is normal, acceptable and enjoyable. The child pornography used for this purpose depicts children who are smiling, laughing and

---

[17] *See* Schrock and Boyd, *Online Threats to Youth: Solicitation, Harassment, and Problematic Content*, *op. cit.*

[18] Ibid.

[19] Ibid.

[20] Ibid.

[21] Ibid.

[22] Ibid.

seemingly having fun, which in turn both legitimizes sex between adults and children and portrays these sexual activities as enjoyable. Of the 1,400 cases of reported child molestation in Louisville, Kentucky, between 1980 and 1984, pornography was connected with every incident and child pornography was connected in the majority of cases.

6.4 Minimizing Risks

For those areas of the world where children have access to the internet, especially where they have access without a requirement of supervision, it is likely that both the positive and negative effects of this accessibility will become ever more evident (Bross 2005, p. 749). Surveys of parents suggest that they buy home computers and subscribe to internet access to provide educational opportunities for their children and to prepare them for the information-age (Subrahmanyam et al. 2001, p. 8). In the U.S.A., a study found that (55 %) of children aged 12–15 stated that they did not tell their parents everything they did on the internet, yet adults kept an eye on children's internet use (91 %), limited online hours (62 %), and used software to filter or block questionable websites (32 %); moreover, two-thirds (67 %) of children surveyed had to ask permission to access the internet (Cankaya et al. 2009, p. 1108). A number of strategies for disrupting the sexual abuse of children using the internet can be imagined. Several of these are addressed by Mitchell et al. (2005). Broadly, they include the use of technology, law, parenting practice, and culture, "behavioral immunization" or "attitude inoculation" with children. There are several kinds of software tools available to families.

"Time-Limiting" software allows parents to set limits on how much time or at what time a child can use the computer or internet. "Filtering and Blocking" software limits access to some sites, words, and/or images. "Outgoing Content Blocking" regulates the content leaving the computer to prevent children from revealing personal information, such as names, addresses and telephone numbers, to people they do not know. "Kid-Oriented Search Engines" work like regular search engines but also provide special features to screen out inappropriate material. Finally, "Monitoring Tool" software informs adults about children's online activity by recording the addresses of visited websites or displaying warning messages to children if they visit inappropriate websites, without necessarily limiting access. Some software incorporates several of these features.

Another illustration of the use of technology to protect minors against online harm is 'Safer Chat', a Belgian initiative of the Minister for Computerization of the State in cooperation with the Belgian Internet Service Providers Association (Lievens 2007, p. 324).

This system requires the use of a child's electronic identity card to gain access to a 'safe' chat room. All children above the age of twelve receive a free card reader in an attempt to promote this feature. In order to verify the age of a person wanting to access a particular 'safe' chat room, the National Registry identification number embedded in the electronic identity card is used (Lievens 2007, p. 324).

In addition to those noted above, other forms of regulation may be adopted, such as the establishment of codes of conduct and contact points (Arnaldo 2001, p. 143). A number of service providers' associations have set up internet sites dedicated to reporting illegal content of a pedophile nature or incitement to racial hatred. These sites are intended to assist victims in finding a contact point at which to file their complaint (Ibid). These contact points either deal with claims directly by establishing contact with the site administrator or by referring victims to competent services.

A final strategy is institutional oversight when children are not in the care of parents. Standards need to be drafted and enforced for libraries, schools, computer camps, and generally in every setting where legal minors are not in the care of their parents but can have unfettered access to the internet (Bross 2005, p. 752). Libraries are, for example, often reluctant to tell users that they cannot access a particular website using the library computer. However, images of sexual contact can then become visible to children as well as other library users, whether wished for or not. Adult book stores and movie houses often limit the advertising of their products in terms of the general public walking past the business establishment, as well as limiting access to minors by law. These are just a few of the possible solutions that individuals and institutions charged with the duty of out of home supervision must consider and act upon (Bross 2005, p. 752).

6.5 Where Next?

In terms of criminal law, controlling cyberspace seems likely to dominate the attention of policy makers and politicians for the foreseeable future. Successful models of enforcement, such as that achieved in the field of child abuse images, are likely to be seized upon by governments and imposed across an ever wider range of issues, resulting in an increasingly mediated access to the internet (Walden 2007, p. 398). Innovative ways of attacking, exploiting and interfering with computer and communication technology are likely to regularly emerge, challenging investigators and prosecutors alike. However, such integrity threats can be expected to attract policy makers only to the extent that they are carried out against, or on a scale, which is perceived to threaten a nation's critical infrastructure. A common concern voiced by digital investigators is that technology has outstripped the national legislation. While we must accept this as an inevitable feature of the landscape, what are the possible consequences? Walden has stated:

> [T]here is the possibility that technology may reach a state, where our use of ICTs no longer provides a valuable source of forensic material for investigators. Data criminals will be able to operate using identities and from locations that are effectively hidden, processing information protected by unbreakable codes. To address this scenario, already part reality, governments may be tempted to give law enforcement agencies further wide—ranging

powers to intrude and disrupt such online criminalities. This may include revisiting ex ante demands that's suppliers of networks and devices design their equipment in a manner supportive of such policing methods.

Accordingly, cybercrime, online child abuse offences and digital investigations will comprise a substantial part of criminal policy, law and practice over the coming years, as information becomes the cornerstone of the global economy.

Acknowledgments An earlier version of this chapter was published in JDFSL 2009 (4). The author is grateful to Glenn Dardick.

References

Y. Akdeniz, *Internet Child Pornography and the Law* (Ashgate, Hampshire, 2008)
C. Arnaldo, *Child Abuse on the Internet: Ending the Silence* (Berghahn Books, NY, 2001)
A. Beech et al., The internet and child sexual offending: a criminological review. Aggress. Violent Behav. **13**, 216–228 (2008)
D. Bross, Minimizing risks to children when they access the world wide web. Child Abuse Negl. **29**, 749–752 (2005)
P. Bocij, *Cyberstalking* (Praeger, Connecticut, 2004)
S. Cankaya, O. Hatice, Parental controls on children's computer & internet use. Procedia Soc. Behav. Sci. **1**, 1105–1109 (2009)
E. Casey, *Digital Evidence and Computer Crime* (Elsevier, NY, 2004)
M. Chawki, *Combattre la Cybercriminalité* (Editions de Saint Amans, Perpignan, 2008)
D. Delmonico et al., *Cybersex Unhooked: A Workbook for Breaking Free of Compulsive Online Sexual Behavior* (Gentle Path Press, Carefree, 2001)
N. Dixon, *Catching Cyber Predators: The Sexual Offences Bill 2002*, (Queensland Parliamentary Library, QLD Brisbane, 2002)
N. Doring, The internet's impact on sexuality: a critical review of 15 years of research. Comput. Hum. Behav. **25**, 1089–1101 (2009)
I. Elliott et al., Understanding online child pornography use: applying sexual theory to internet offenders. Aggress. Violent Behav. **14**, 180–193 (2009)
M. Ferraro et al., *Investigating Child Exploitation & Pornography: The Internet Law and Forensic Science* (Elsevier, NY, 2005)
Garlik, UK Cybercrime Report, Online (2008)
T. Hoeren, The European Liability and Responsibility of Providers of Online Platforms such as Second Life, JILT 1 (2009)
Internet Watch Foundation, *Annual Report on Cybercrime*(Cambridge University Press, Cambridge, 2008)
S. Kierkeggard, Cybering, Online Grooming, and Age Play. Computer Law and Security Report **24**, 41–55 (2008)
E. Lievens, Protecting children in the new media environment: rising to the regulatory challenge? Telematics Inform. **24**, 315–330 (2007)
I. Marsh, M. Gaynor, *Crime, Justice and the Media* (Routledge, NY, 2009)
K. Mitchell et al., Protecting youth online: family use of filtering and blocking software. Child Abuse Negl. **29**, 753–765 (2005)
L. Penna et al., *Challenges of Automating the Detection of Pedophile Activity on the Internet*, The First International Workshop on Systematic Approaches to Digital Forensic Engineering (SADFE'O5), November 7–9, Taipei (2005)

C. Preston, All knowledge is not equal: facilitating children's access to knowledge by making the internet safer. Int. J. Commun. Law Policy **13**, 115–132 (2009)

E. Quayle et al., In: Child pornography and sexual exploitation of children online. The World Congress III against Sexual Exploitation of Children and Adolescents, Rio de Janerio 25–28 November, 2008

B. Schell et al., Cyber child pornography: a review paper of the social & legal issues and remedies —and a proposed technological solution. Aggression Violent Behav. **12**, 45–63 (2007)

K. Sheldon, D. Howitt, *Sex Offenders and the Internet* (John Wiley, West Sussex, 2007)

A. Schrock et al., *Online Threats to Youth: Solicitation, Harassment, and Problematic Content* (Berkman Center for Internet & Society, Cambridge)

P. Sommer, *Evidence in Internet Pedophilia Cases*, NCS/ ACPO Conference, July 2002, Bournemouth

K. Subrahmanyam et al., The impact of computer use on children's and adolescent's development. Appl. Dev. Psychol. **22**, 7–30 (2001)

D. Taylor, Internet Service Providers (ISPs) and their Responsibility for Content under New French Legal Regime. Comput. Law Secur. Rep. **20**(4), 268–272 (2004)

D. Van der Merwe et al., *Information and Communications Technology Law* (Lexis Nexis, Durban, 2008)

I. Walden, *Computer Crimes and Digital Investigations* (Oxford University Press, Oxford, 2007)

D. Wall, *Crime and the Internet* (Routledge, Oxford shire, 2001)

D. Wall, *Legal professionalism in the information age*. The Paralegal Educ. **22**(1), 50–54 (2008)

S.-J. Wang, Measures of retaining digital evidence to prosecute computer—based cybercrimes. Comput. Stan. Interfaces **29**(2), 216–223 (2007)

J. Wolak et al., Predators and their victims. Am. Psychol. Assoc. J. **63**(2), 111–128 (2008)

R. Wortley, *Child Pornography on the Internet*, (US Department of Justice, Washington, 2006)

Part V
Privacy, Security and Crime Control

Chapter 7
Anonymity, Privacy and Security Issues in Cyberworld

7.1 Introduction

Anonymity, often considered a cornerstone of democracy and a First Amendment guarantee, is easier to attain than ever before, due to the emergence of cyberspace (du Pont 2001, p. 191). Cyberspace enables people to share ideas over great distances and engage in the creation of an entirely new, diverse, and chaotic democracy, free from geographic and physical constraints (N'oeil 2001). As of September 30, 2009, about 1,733,993,741 users had access to the internet. Those numbers are growing rapidly. Due to the nature of ICTs, identities in cyberspace are easily cloaked in anonymity (du Pont 2001, p. 192). Once a message sender's identity is anonymous, cyberspace provides the means to perpetrate widespread criminal activity among the masses, with little chance of being apprehended (du Pont 2001, p. 192; Brenner and Goodman 2002). In a report to former Vice President Al Gore, Attorney General Reno found a need for greater control of anonymity in cybercrime. Reacting to several attacks on eBAY, CNN.com and other websites, former President Clinton underscored the opinion that the government needs to maintain a watchful eye on cyberspace. On the other hand, anonymity in cyberspace allows whistle-blowers and political activists to express opinions critical of employers and the government, enables entrepreneurs to acquire and share technical information without alerting their competitors, and permits individuals to express their views online without fear of reprisals and public hostility (Aldesco 2002). It is clear that in various parts of the world, people may have an interest in not being identified and thus connected to certain published views and opinions (Nicoll et al. 2003). Due to the international character of the internet, those reasons for anonymous communications which are related to the "freedom of expression" may gain new dimensions (Lipschultz 2000).

Before the information age, a person's identity, and information relating to his or her identification seemed to be more precisely controlled (Nicoll et al. 2003, p. 3).

M. Chawki et al., *Cybercrime, Digital Forensics and Jurisdiction*,
Studies in Computational Intelligence 593, DOI 10.1007/978-3-319-15150-2_7

But all that has changed. The advent of the information society has vastly increased the need for identifying mechanisms and thus, public availability of the relevant technologies (Nicoll et al. 2003, p. 3). Names, addresses, e-mail addresses, photographs, social security numbers, etc., are freely available on the internet and numerous identity related characteristics are for sale (Nicoll et al. 2003, p. 3). On the internet, anyone has the opportunity to gain knowledge about other people. The development of ICTs makes more and more people reluctant to reveal their true identity (Nicoll et al. 2003, p. 3).

In combination with this, different services have recently been developed which make internet activities, such as surfing, anonymous. Facilities are also anonymous. Facilities are also available to provide individuals with a pseudo identity (Nicoll et al. 2003, p. 3). Hence, anonymous communication is promoted as the solution to the problem. Anonymity does, however, raise various legal questions: What exactly do we mean by anonymity? Why would people want to communicate and transact on an anonymous basis? What are the practical and legal constraints upon anonymity when communicating and transacting with others? Finally, total anonymity may be possible through the use of privacy-enhancing technologies.

7.2 Anonymity in Cyberspace

There are two kinds of anonymity: true anonymity and pseudo-anonymity (du Pont 2001, p. 192). However, some scholars fail to sufficiently address this distinction. Dialogue on the issue of anonymity legislation suffers on account of this lack of distinction. This section will therefore distinguish between true and pseudo—anonymity, two completely different forms of expression, with differing degrees of political and social value and constitutional protection (du Pont 2001, p. 192).

7.2.1 True Anonymity

Truly anonymous communication is untraceable. Indeed, only coincidence or purposeful self-exposure will bring the identity of the mystery sender to others; the identity of a person acting in a truly anonymous manner cannot be definitively discovered with any amount of diligence (du Pont 2001, p. 192). Attempts can be made to discover the identity of the sender through inference, but any concrete trail of clues betraying the message sender has been erased by circumstance, the passage of time, or by the sender herself. Although some forms of truly anonymous communication, such as political speech, are considered valuable, this form of anonymity has exceptional potential for abuse because the message senders cannot be held accountable for their actions (du Pont 2001, p. 192).

7.2.2 Pseudo-Anonymity

Pseudo-anonymous communication is inherently traceable (Rigby 1995). Though the identity of the pseudo-anonymous message sender may seem truly anonymous because it is not easily uncovered or made readily available by definition, it is possible to discover their identity. This kind of anonymity has significant social benefits; it enables citizens of a democracy to voice their opinions without fear of retaliation against their personal reputations, but it forces them to take ultimate responsibility for their actions, should the need somehow arise (Rigby 1995). Although governments could abuse their ability to uncover the identity of people acting pseudo—anonymously, it is not in the government's interest to break that trust; by respecting pseudo-anonymous identities, governments can often avoid the far more dangerous abuses stemming from true anonymity (Rigby 1995).

7.2.3 Anonymity, Privacy and Freedom of Speech

The world in which we live can frequently be extremely conservative, often making it dangerous to make certain statements, have certain opinions, or adopt a certain lifestyle (Brin 1994; Rigby 1995). Anonymity is important for online discussions involving sexual abuse, minority issues, harassment, sex lives, and many other things. Moreover, anonymity is useful for people who want to ask technical questions that they do not want to admit they do not know the answer to, report illegal activities without fear of retribution, and many other things. Without anonymity, these actions can result in public ridicule or censure, physical injury, loss of employment or status, and in some cases, even legal action. Protection from harm resulting from this type of social intolerance is a definite example of an important and legitimate use of anonymity on the internet. An example of how vital such anonymity can be is exemplified by the following excerpt from a newsgroup post during a temporary shutdown of penet.fi (Rigby 1995):

> I had been posting to a non-technical *misc.* newsgroup about an intimate topic for which I felt I required privacy. I have received immeasurable help from the people in that news group...Please, folks, believe me, I *need* this service. Please consider my point of view and permit admin@penet.fi to turn the service back on.

Doctors who are members of the online community, moreover, usually encourage their patients to connect with others and form support groups on issues about which they do not feel comfortable when speaking publicly. It is important to express certain opinions without revealing our true identities. One relevant example of anonymity in the physical space is the debate over Caller ID on telephones. A great number of people were extremely disturbed that the person at the receiving end of a telephone call would know the identity of the caller. People had taken it for granted that they could be anonymous if they wanted and were distressed at the idea of that anonymity being taken away. Many Net users feel the same way about online anonymity.

Anonymity allows an individual to seek online information, resources and support without jeopardizing their public reputation and relationships. Fear of discrimination might prevent an individual from seeking help. Anonymity allows information gathering about issues like addiction to alcohol, gambling, drugs or sex; sexual identity, where identifying as non-heterosexual could cause problems at work or home; testing or treatment options for illnesses like AIDS; or information about birth control or sexually transmitted infections. Especially for youth, the ability to search anonymously allows individuals to make informed and responsible choices with information they may not have been able or willing to locate had they been forced to disclose their identities.

Society benefits when people are not afraid to seek help. Individuals have the right to be left alone, in this case to not be persecuted for challenging societal norms and this is facilitated by the opportunity to operate anonymously in cyberspace. Anonymity is effective in promoting freedom of expression. Julf Helsingius asserts that anonymity is beneficial because it gives people an outlet for their opinions, even controversial ones. He argues that it is "good to bring out things like that in daylight because that actually allows you to … start processing it, see how people react to it, and so on" (Rigby 1995).

This may have sort of a cathartic effect in that it allows people to get their feelings out without physically hurting people of other cultures, races, etc. In addition to this, anonymity hinders some methods of controlling the actions of other people. This is a further argument for the usefulness of anonymity in the protection of freedom of expression. There are many long-standing precedents for anonymity in publishing. The responsibility of journalists not to reveal their sources is recognized almost universally. Many authors write under pen-names and there are still some cases where the true identity of the author has never been discovered. Even the Federalist Papers were published under a pseudonym. Most newspapers publish letters to the editor and help columns and allow the letters to be anonymous or signed with a pseudonym and many newspaper papers are merely credited to 'AP Newswire'. Additionally, anonymous peer reviews of proposals and papers are common in academic circles (Rigby 1995).

In *Forensic Advisors v. Matrix Initiatives*, a Maryland Court has been asked to order the disclosure of the identity of subscribers to a newsletter. In this affair, *Matrix*, a pharmaceutical company, was seeking a publisher's subscriber list to use in connection with a lawsuit filed against numerous individuals who posted allegedly derogatory comments about the company to an internet discussion board. Privacy and civil liberties advocates, including EPIC, the Electronic Privacy Information Center have filed an amicus brief in support of the publisher's right to protect the list. The brief argues that the list ought to be protected under Maryland law that protects a journalist's sources. They further argue that subscribers have a right to remain anonymous under the US Constitution's First Amendment, since disclosure of the list would deter readership and violate established privacy rights. In *ACLU v. Miller*, the American Civil Liberties Union got an injunction against the enforcement of a Georgia statute that prohibited a person from falsely identifying herself while sending e-mail, posting on the internet, and more (one of the problems with the statute was that it was too vague). The court ruled that it was appropriate to grant an injunction, among other reasons, when there was the potential for chilling free expression. The court agreed with the state that its purpose in enacting the statute—preventing fraud—was a compelling state interest, but decided to side against the state because the statute was not narrowly-enough tailored to its purpose.

Finally, anonymous communication can be achieved in real life by sending an unsigned letter or making an anonymous phone call. From the large number of users who take advantage of anonymous services on the internet, it can be seen that these services are genuinely necessary and fulfill a specific need. The availability of the technology to set up such an anonymous server also makes the elimination of such servers virtually impossible; as soon as one is shut down, another one is created. The current availability of such services eliminates the need to forge an identity or use another person's identity to correspond anonymously. People on the Net are anonymous to some degree anyway because of the inherent characteristics of the medium. Services providing additional anonymity are only expanding on this feature of the internet. Pseudonymity therefore comes in useful in that it allows users to send mail to pseudonymous users in response to their mail or post. People are able to respond to emails that they like or dislike or that they find offensive or disruptive. This makes the pseudonymous user more responsible for his or her actions than the completely anonymous user. They are still accountable for their online actions, but are protected from 'real world' damage.

Abolishing anonymity servers is not necessary since the technology exists to produce kill files which allow users to choose for themselves what they consider offensive. This allows individuals to filter out anonymous posts and emails which they dislike, while still reaping the benefits afforded by anonymous services. Although some people will automatically discount any anonymous postings, other people do not care who wrote it, as long as it is intelligent or funny. Still others use anonymity specifically to allow their opinions to be judged on their merit, rather than by the name attached to them.

7.3 Impact and Harm Generated by Anonymity

Although anonymity is extremely important for the protection of human rights, it is also tied to cybercrime, and it is claimed that it allows criminals to use the internet without the possibility of detection (Akdeniz 2002). With respect to cyberspace, identifying an electronic crime scene can be a daunting task when the perpetrator may have routed his communications with the victim through computers in three or four countries, with obscure networks that are inaccessible to investigators. Additionally, perpetrators could make things much more difficult and complicated by using technology and encryption techniques that provide a high level of anonymity or assuming the identity of an innocent person. Moreover, the scale of cybercrime can exceed that of real-world crime in terms of the degree of harm inflicted by a single crime. In modern times there is more focus on protecting the 'container' of valuables (the computer is merely the modern equivalent of a bank vault, only instead of money or gold it contains data) than protecting the real valuables in most ICT crimes, namely the data contained in the computer, the cell phone's GPS device and so on (Snail 2009).

Criminals who wish to use a computer as a tool to facilitate unlawful activity may find that the internet provides a vast, inexpensive and potentially anonymous way to commit unlawful acts, such as fraud, the sale or distribution of child pornography, the sale of guns, drugs or other regulated substances without regulatory protection, and the unlawful distribution of computer software or other creative material protected by intellectual property rights (The Electronic Frontier 2000). For example, some services like anonymous re-mailers can plainly frustrate legitimate law enforcement efforts despite providing privacy and encouraging freedom of expression. In the first case to be prosecuted in Queensland, a woman received e-mail correspondence that began amicably, but then became more threatening once she sought to end the communications. She ultimately received death threats from the offender and threats to have her gang raped, the scene videotaped and uploaded on the internet. In another case brought to court in the USA, a University student harassed five female students after buying information about them via the net. The student sent over 100 messages including death threats, graphic sexual descriptions and references to their daily activities. In another recent case, a phisher e-mail claiming to be from MSN was sent to computer users. It said:

"We regret to inform you that technical difficulties arose with our recent update. Unfortunately part of our customer data base and backup system became inactive." This authentic-looking message offered a toll free telephone number in addition to a web link and urged individuals to click on the link to the phoney website. The message then informed individuals that they needed to enter their personal information. Later on, they realized that they had been victims of a phishing attack.

Thus, territorial-based strategies tend not to be effective against online anonymity because they are designed to prevent the citizens of one nation-state from preying on each other, not to prevent their preying on citizens of other nation-states (Brenner 2004). In this respect, Marc Goodman has succinctly stated (Goodman and Brenner 2002):

> [L]aw has evolved to maintain order within a society. Each nation-state is concerned with fulfilling its obligations to its citizens … [N]o nation can survive if its citizens are free to prey upon each other. But what if they prey upon citizens of another society? What if the citizens of Nation A use cyberspace to prey upon the citizens of Nations B and C? Is this a matter that is likely to be of great concern to Nation A? There are historical precedents for this type of behavior that may shed some light on what will ensue in cyberspace. The most analogous involves high-seas piracy and intellectual piracy. Both involved instances in which societies were willing to allow (or even encourage) their citizens to steal from citizens of other societies. In both, the focus was on crimes against property and the motivation was purely economic. [T]he conduct took place at the 'margins' of the law: high-seas piracy occurred outside the territorial boundaries of any nation and therefore outside the scope of any laws; 18th-Century American intellectual property piracy (Brenner 2004) occurred when the legal

status of intellectual property as 'property' was still evolving. Both were outlawed when they became economically disadvantageous for the host countries. One can, therefore, hypothesise that countries may be inclined to tolerate their citizens' victimizing citizens of other nations if (a) the conduct takes place at the margins of the law and (b) it results in a benefit to the victimizing nation. The former gives the victimizing nation plausible deniability when confronted with its tolerance of illegal activity; the latter is an obvious motive for tolerating the activity.

Accordingly, law enforcement agencies are faced with the need to evaluate and determine the source, typically on very short notice, of anonymous e-mails that contain bomb threats against a given building or threats to cause serious bodily injury (Levy 2005). Thus, internet based activities should consistently, with physical world activities and in a technology-neutral way, further important societal goals (such as the deterrence and punishment of those who commit money laundering). National policies concerning anonymity and accountability on the internet thus need to be developed in a way that takes account of privacy, authentication, and public safety concerns (Levy 2005). In one recent case, Judy McDonough, a 56 years-old occupational psychologist from Shaw, England, suffered a disturbing blow when she realized that someone had stolen her identity from the internet (Levy 2005). By that time, however, the thief had already opened two credit cards in her name, taken out three bank loans and ordered £23,000 in debt in 3 years (Levy 2005). McDonough tried six times to report the crime to the local authorities, but bank officers made lackluster efforts to help. Finally, McDonough turned to her Member of Parliament for assistance. Hitherto, the thief—who McDonough suspects is a relative, has not been caught (Levy 2005). In another case (Privacy Rights Clearing House (UCAN) 2005), an American citizen tried to sell his house in California. He contacted several real estate agents to discuss with them a listing for the house. He was then informed by these agents that his house has been rented to individuals that he was not aware of or has even agreed to rent his house to. Someone was collecting the rent on his house, and upon checking with the US county records, he found out that someone had not only used his name and arranged to fake his signature, but made a power of attorney document in his name to received loans on his property, bought a business in his name and accumulated a huge amount of financial burden in his name as well. The personal information of this victim was found and downloaded from internet.

7.4 Regulating Anonymity in Cyberspace

From logical, theoretical, and pragmatic perspectives, knowing the problem, the risks associated therewith, and the ills resulting therefrom is an indispensable step towards a possible regulation. Since these issues are difficult and sensitive, it is not easy to decide how to legally regulate anonymity in cyberspace. According to an EC report, published in 1999 (Privacy Rights Clearing House—UCAN):

> Users may wish to access data and browse anonymously so that their personal details cannot be recorded and used without their knowledge. Content providers on the internet may wish to remain anonymous for legitimate purposes, such as where a victim of a sexual offence or a person suffering from a dependency such as alcohol or drugs, a disease or a disability wishes to share experiences with others without revealing their identity, or where a person wishes to report a crime without fear of retaliation. A user should not be required to justify anonymous use. Anonymity may however also be used by those engaged in illegal acts to complicate the task of the police in identifying and apprehending the person responsible. Further examination is required of the conditions under which measures to identify criminals for law enforcement purposes can be achieved in the same way as in the 'off-line' world. Precedents exist in laws establishing conditions and procedures for tapping and listening into telephone calls. Anonymity should not be used as a cloak to protect criminals.

At present, there is no consistent policy which can be discerned in any one jurisdiction that would allow the resolution of the tensions illustrated above. Each problem relies on striking a fair balance between the interests of the individual on the one hand, and the interests of the State on the other. Various countries have laws both protecting and forbidding anonymity. For example, many countries have laws protecting the anonymity of a person giving tips to a newspaper, and laws protecting the anonymity in communication with priests, doctors, etc. (Palme and Berglund 2002). On the other hand, the obvious risk of misuse of anonymity has caused some countries to try special legislations concerning its regulation. Cases of defamation often result in corporations seeking motions to uncover the identities of individuals who have made negative comments on bulletin boards or websites (Stiles 2002). Although hurtful, these comments are often opinions, not facts, and, therefore, not punishable crimes (Stiles 2002). In the case of cyber-trespass, it is first required that plaintiffs show damages caused by defendants. Safeguards ensure that anonymity is protected until proof of a crime exists. These safeguards prevent an ISP from providing a "subscriber's personal information without the subscriber's knowledge and consent, except in certain specified circumstances" (Homsi and Kaplan 2005). Accordingly, the Council of European Union has adopted a Directive of the European Parliament and the Council on data retention (Directive on Data Retention, 2005/0182/COD), amending directive 2002/58/EC. The Directive aims to harmonize member states' provisions concerning the obligations of the providers of publicly available electronic communications service or of public communications networks with respect to the retention of certain data which are generated or processed by them, in order to insure that the data are available for the purpose of the investigation, detection and prosecution of serious crime, as defined by each member state in its national law.

The Directive is applied to traffic and location data on both legal entities and natural persons, and to the related data necessary to identify the subscriber or registered user. It shall not be applied to the content of electronic communications; including information consulted using an electronic communications network. The data retained is made available only to the competent national authorities in specific cases and in accordance with national law. It is retained for periods of no less than 6 months and no more than 2 years from the date of communication. "Agreement on retaining communications data places a vital tool against terrorism and serious crime in the hands of law enforcement agencies across Europe."

Member states have to take necessary measures to ensure that any intentional access to, or transfer of, retained data is punishable by penalties, including administrative or criminal penalties that are effective, proportionate and dissuasive. Each member state will designate a public authority to be responsible for monitoring the application within its territory of the provisions adopted regarding the security of sorted data. At the same time, governments have confronted the dangers of cyberspace by devoting significant resources towards formulating a legal framework that addresses the technical and operational challenges of crime. The Convention on Cybercrime is considered "one of the most important legal instruments elaborated within the Council of Europe". It was approved by the Committee of Ministers of the Council of Europe (COE), and on November 23, 2001, the Convention was signed by 26 member states of the COE along with four non-member states—Canada, Japan, South Africa, and the USA, and came into force on July 7, 2004. The Convention is the first international treaty to allow police in one country to request that their counterparts abroad collect an individual's computer data, have the individual arrested and extradited to serve a prison sentence abroad (Godwin 2001). It is aimed principally at:

- harmonizing the domestic criminal substantive law elements of offences and connected provisions in the area of cyber-crime;
- providing for domestic criminal procedural law powers necessary for the investigation and prosecution of such offences as well as other offences committed by means of a computer system or evidence in relation to which is in electronic form;
- setting up a fast and effective regime of international cooperation.

The Convention defines substantive criminal laws to be legislatively adopted by all signatory states. It covers crimes in four main categories:

- offences against the confidentiality, integrity and availability of computer data and systems (Cybercrime Convention, Articles 2–6)
- computer-related offences (Cybercrime Convention, Articles 7 and 8)
- content-related offences (for example, child pornography) (Cybercrime Convention, Article 9)
- offences related to infringements of copyright and related rights (Cybercrime Convention, Article 10).

The Convention also seeks to harmonize new procedures and rules of 'mutual assistance' to aid law enforcement in the investigation of cybercrimes. Signatory countries are required to ensure that certain measures are available under their national law: "[e]expedited preservation of stored computer data", expedited preservation and disclosure of traffic data; the ability to order a person to provide computer data and to order an ISP to provide subscriber data under its control; "[r] eal-time collection of traffic data" (Cybercrime Convention, Article 20) and interception of content data (Cybercrime Convention, Article 21). The Convention provides that signatory countries must adopt measures to establish jurisdiction over any offences committed in their respective territories or by their nationals (Cybercrime Convention, Article 22). Moreover, it empowers legal authorities and police in one country to collect evidence of cybercrimes for police in another country, and establishes a '24/7 network' (Cybercrime Convention, Article 35) operating around the clock, 7 days per week, to provide immediate assistance with ongoing investigations.

According to Article 15, which deals with 'conditions and safeguards', the "establishment, implementation and application of the powers and procedures provided for in Section 2 of the Convention [pertaining to procedural law] are subject to conditions and safeguards" provided under the domestic law of each signatory country. These domestic laws and procedures shall include conditions or safeguards, which may be provided constitutionally, legislatively, judicially or otherwise. The modalities should include the addition of certain elements as conditions or safeguards that balance the requirements of law enforcement with the protection of human rights and liberties. As the Convention applies to parties of many different legal systems and cultures, it is not possible to specify in detail the applicable conditions and safeguards for each power or procedure (Convention on Cybercrime, Explanatory Report). Parties should ensure that these conditions and safeguards provide for the adequate protection of human rights and liberties. There are some common standards or minimum safeguards to which parties to the Convention must adhere. These include standards or minimum safeguards arising pursuant to obligations that a party has undertaken under applicable international human rights instruments (Convention on Cybercrime, Explanatory Report). These instruments include the 1950 European Convention for the Protection of Human Rights and Fundamental Freedoms and its additional Protocols Nos. 1, 4, 6, 7 and 12 (ETS Nos. 005, 009, 046, 114, 117 and 177), in respect of European States that are party to them. It also includes other applicable human rights instruments in respect of states in other regions of the world (e.g., the 1969 American Convention on Human Rights and the 1981 African Charter on Human Rights and Peoples' Rights) which are party to these instruments, as well as the more universally ratified 1966 International Covenant on Civil and Political Rights. In addition, there are similar protections provided under the laws of most states (Convention on Cybercrime, Explanatory Report).

Article 19 of this Convention aims at modernising and harmonising domestic laws on the search and seizure of stored computer data for the purposes of obtaining evidence with respect to specific criminal investigations or proceedings (Cybercrime Convention, Article 19). Any domestic criminal procedural law includes powers for the search and seizure of tangible objects. However, in a number of jurisdictions, stored computer data per se will not be considered a tangible object, and therefore cannot be secured on behalf of criminal investigations and proceedings in a parallel manner to tangible objects, other than by securing the data medium upon which it is stored (Cybercrime Convention, Article 19).

To strike a balance between privacy and security in cyberspace, Article 8 of the European Convention on Human Rights and Fundamental Freedoms denotes a right to *respect* for a citizen's private life, his home and his correspondence. But the flexible notion of 'respect' is informed by the interests of national security, public safety, the economic well-being of the country, prevention of disorder and crime (Nicoll et al. 2003), protection of public morals and the rights and freedoms of others. In the USA for example, there is no express right to privacy guaranteed by the Constitution. There, the recognition of the need for citizens to be able to communicate anonymously is derived from the right to speak freely, to dissent and criticize (Nicoll et al. 2003). Clear guidelines seem only to exist in the political context. Commercial interests are accorded reduced protection. There is no equivalent of the express reference to the economic well-being of the country that exists in the European Convention on Human Rights and Fundamental Freedoms (Nicoll et al. 2003). It may be that the market will regulate itself. That may involve tough compromises or the 'spending' of privacy which becomes tantamount to an asset. The countervailing benefit is some form of financial gain. The market may also regulate itself through a professional body or association. For example, 'Yahoo!' has policies that allow it to reveal the identities of its users when the service provider is subpoenaed, subjected to court orders or involved in a legal process (Stiles 2002). The fact that these companies can, and will, identify internet users when asked by a court to do so, forces courts to decide whether to protect the anonymity of internet users sued for crimes or require their identity to be revealed in order to have a more easily administered crime lawsuit. But here, as illustrated by Carr (Stiles 2002), in connection with the internet Watch Foundation, a private association may effectively block the door to the internet or restrict permissible activities in the absence of any public debate or even in direct opposition to public demand. The danger of this sort of private intervention is that it may often occur because the trade body concerned fears government regulation. The government is able to abdicate its responsibilities in a politically unproductive or dangerous area by permitting a form of delegated legislation while avoiding any accountability (Nicoll et al. 2003).

Yet, fundamental to the regulation of anonymous internet activity is the recognition that communication is not 'geographically contained'. The nature of the medium dictates that the prevention of cybercrime, for example, must be accompanied by a degree of international cooperation that has not so far been possible to attain in other contexts. Not only is it difficult, owing to political and cultural differences, to reach an international consensus on a list of alleged crimes that would justify a coordinated approach in their detection, but the process is further exacerbated by wide dissemination of evidence, the transient nature of much of the evidence and a trail that quickly turns cold. According to Sims (Nicoll et al. 2003), various procedural means are available in common law countries to gain the courts' assistance in breaking through identity barriers.

These methods can be hampered however, by a lack of formalized transitional cooperation. The nature of cyberspace is not only a problem in securing a uniform approach to online anonymity. Howells and Edwards argue that anonymity gives an unfair advantage to lobby or interest groups who are able to band together and thereby to focus attention on influencing legal developments to their advantage, at the expense of less cohesive or numerically manageable interests such as consumers. Ironically, it is consumers who are the major driving forces in the growth of e-commerce. Yet, surveys have shown that they have little confidence in the medium, an attitude that is perhaps disproportionately affected by invasions of privacy such as spam and junk mail which, whilst they do little economic harm, can cause huge annoyance.

7.5 The Case Law

In the last years, some countries have seen the internet as a threat to their way of life because of how easy it is for people to share ideas and opinions. They may have rules that prevent their citizens from using the internet, limit which websites they are allowed to visit, or prevent them from sharing their own opinions in cyberspace.

7.5.1 Al Farhan Case

In December 2007, Saudi Arabia's most popular blogger, Fouad al-Farhan, was detained for questioning because he criticized the government on his blog (The Washington Post, January 2008). This was the first known arrest of an online critic in the Kingdom. Farhan used his blog to criticise corruption and call for political reform, and thus was detained for violating rules not related to state security. Farhan has had trouble with the authorities in KSA before. In 2006, he was told by the Interior Ministry to tone down his blog and to dissolve an association he was forming to protect bloggers' rights. He dissolved the group and quit blogging for 9 months because of his business interests, but then later went back to blogging even more critically (Ibid).

7.5.2 Bin Saleh Case

In January 2009, Saudi authorities detained Hamoud Bin Saleh and blocked his blog due to his overt expression of his opinions, and the announcement on his blog that he had converted from Islam to Christianity (IFEX Website). Hamoud Bin Saleh, a 28 year old Saudi national, was arrested after posting comments on his blog describing his conversion to Christianity and criticizing the Saudi judicial system. Several years ago, after Hamoud witnessed three Pakistanis convicted on drug charges being publicly beheaded, he became disillusioned with Islamic justice and began researching comparative religions while travelling and studying abroad. As a result of his study, he converted to Christianity and began advocating equal rights for Christians in Saudi Arabia. Hamoud had been arrested previously for 9 months in 2004 and for a month in November 2008, at which time he was released for the duration of a Saudi-sponsored interfaith conference in New York and immediately re-arrested. Hamoud's blog has now been blocked by Saudi authorities.

7.6 The Appropriate Balance

As we move into the computerized world of the 21st century, it is being argued that the right to privacy is an important fundamental right for the user-citizen. This is not just about hiding data, but about self-possession, autonomy, and integrity. The issue of privacy on the internet includes user anonymity, cyber harassment and violation of expectation of privacy in surveillance. In United States v. Mark L. Simons, it was held that employees are not entitled to have a reasonable expectation of privacy when it comes to their work related electronic communications. In United States v. William W. Ziegler, the Court acknowledged that an employee has a right to privacy in his workplace computer. However, it also found that an employer can monitor illegal activities.

The two sides in this chapter seem to be irreconcilable. Is there a middle ground? The European Union directives on privacy protection are comprehensive and protect individual's access and control of the use of personal information. These Directives also ensure that the information is accurate, complete and recent, and emphasize 'safe harbor' solutions. 250 years ago, Benjamin Franklin summarized the tension between anonymity and security in the following words: "Those who can give up essential liberty to obtain a little temporary safety deserve neither liberty nor safety". But Franklin had never heard of Osama bin Laden, and he could not have even begun to imagine the use of airplanes as weapons of mass-murder by stateless martyrs. In 1963, Supreme Court Justice Arthur Goldberg provided another apt summary that captures the need for the balance we see developing: "While the Constitution protects against invasions of individual rights, it is not a suicide pact". If internet anonymity threatened nothing before but dogs pretending to be humans, today internet users ask them to balance Franklin's idealism with Goldberg's realism. We will thus continue to search for that appropriate point of balance.

7.7 Summary

This chapter has explored and analyzed anonymity in cyberspace. It demonstrates that there are no definite guidelines to determine the boundaries of anonymity and interests that show whether, and to what extent, limitations on anonymity are required, or not. Furthermore, it also shows that limitations on anonymity could be applied to reflect the legislator's recognition of various interests in making a person's identity known. Accordingly, one of the digital applications that could hold potential for balancing anonymity and the quest of governments and businesses to have identification data available is the facility of Trusted Third Parties, such as the certification authorities or anonymity software. These could play an intermediate role in keeping a true identity secret and also in providing identity and tracing information once certain conditions are satisfied. There will be always be a continued debate, such as in France, whether they must retain in escrow, identifying information in the event that governments require to decrypt messages, ostensibly for state security reasons. In line with the present developments in the USA, where internet service providers have to reveal the identity of people posting information through their facilities, case law, self-regulatory initiatives and maybe even legislation may set the conditions under which identifying information must be revealed by intermediaries.

Acknowledgment A version of this chapter was published in Complex 3/06—Kierkegaard (2006).

References

Y. Akdeniz, Anonymity, democracy and cyberspace. Soc. Res. **69**(1), 223–237 (2002)

A. Aldesco, The demise of anonymity, a constitutional challenge to the convention on cybercrime. Loy. L. A. Ent. Law Rev. **23**, 81 (2002)

S. Brenner, Cybercrime metrics: old wine, new bottles. Va. J. L. Technol. **9**(13) (2004)

S. Brenner, M. Goodman, The emerging consensus on criminal conduct in cyberspace. UCLA J. L. Tech. **3**, 56–57 (2002)

D. Brin, *The Transparent Society: Will Technology Force Us to Choose between Privacy and Freedom* (Perseus Publishers, New York, 1994)

G. du Pont, The criminalization of true anonymity in cyberspace. Mich. Telecomm. Tech. L. Rev. **7**, 191–216 (2001)

M. Godwin, *International Treaty on Cybercrime Poses Burden on High –Tech Companies*. IP Worldwide (4 April 2001)

M. Homsi, A. Kaplan, Online anonymity and john doe lawsuits (2005), http://www.cippic.ca

S. Kierkegaard, *Legal, Privacy and Security Issues in Information Technology*. Complex (Computer Lex), vol. 1, 03/06

S. Levy, Grand theft identity (Newsweek, 4 July 2005), http://www.newsweek.com. Accessed 03 July 2011

J. Lipschultz, *Free Expression in the Age of the Internet: Social and Legal Boundaries* (West View Press, Oxford, 2000)

C. Nicoll et al., *Digital Anonymity and Law: Tensions and Dimensions* (T.M.C. Press, The Hague, 2003)

R. N'oeil, *The First Amendment and Civil Liability* (Indiana University Press, Bluemington, 2001)

J. Palme, M. Berglund, Anonymity on the internet (2002), http://www.people.dsv.su.se. Accessed 05 Aug 2011

K. Rigby, *Anonymity on the Internet must be Protected, Ethics and Law on the Electronic Frontier* (Fall, Massachusetts Institute of Technology, 1995), http://groups.csail.mit.edu/mac/classes/6.805/student-papers/fall95-papers/rigby-anonymity.html. Accessed 09 Aug 2011

S. Snail, Cybercrime in South Africa—hacking, cracking and other unlawful online activities, J. Inf. L. Technol. 1, (2009)

A. Stiles, Everyone's a critic: defamation and anonymity on the internet. Duke L. Tech. Rev. 0004 (2002)

Chapter 8
Strategies and Statutes for Prevention of Cybercrime

8.1 Introduction

Cybercrime is a global phenomenon, which countering requires global legal strategies. Only by harmonizing national laws and by formalizing countries' mutual cooperation can legal enforcement agencies properly respond to sophisticated, agile methods used by cybercriminals (de Almendia 2011). The rapidly increasing number and volume of cybercrime incidents and losses indicate urgent need of convergence by the international community towards a common set of substantive and procedural legal rules.

However, some issues seem to make it difficult to reach a consensus on the contents of such legal rules. For instance, many countries have not yet framed a balance between security and privacy concerns, thus delaying the process of approving procedural rules. Some other countries resist against joining any international convention which they have not negotiated since its inception, or where they do not have equal opportunity for voicing views or claims.

The outcome has been that a large number of countries still lack adequate cybercrime legislation and/or have not yet acceded to existing relevant Budapest Convention.

Given such a picture, regulation should be improved in order to facilitate more responsive action by the international community in putting together forces against cybercrime. Perhaps the combined use of legal rules and improved information security standards might be an interesting, potentially effective way to address the above referenced issues.

M. Chawki et al., *Cybercrime, Digital Forensics and Jurisdiction*,
Studies in Computational Intelligence 593, DOI 10.1007/978-3-319-15150-2_8

8.2 National and Regional Strategies: The European Approach

Computer crime dates from the origination of computers (Shinder 2002). The first empirical computer crime studies applying scientific research methods were conducted in the 1970s (Parker 1998). These studies verified a limited number of cases and suggested that many more have gone undetected or unreported (Goodman and Brenner 2002). In the United States, the Senator Abraham Ribicoff introduced the first proposed federal computer crime legislation in 1977: the Federal Computer Systems Protection Act. The bill was revised and reintroduced 2 years later (Ibid). It then died in committee (Griffith 1990). However it was influential in promoting the subsequent enactment of federal computer crime legislation and in encouraging the adoption of such legislation in Florida and Arizona (Hogge et al. 2001).

Since then, many new crimoids have emerged. Some crimoids, such as eavesdropping on the radio waves that emanate from computers, have never been proven (Parker 1998). Reports of computer codes, including the Michelangelo and fictitious Good Times viruses, have added to the folklore of computer crimoids (Ibid). The vulnerabilities of the information society and the limitations of the existing computer security approach, as well as legislations and law enforcement efforts, became apparent and widely and publicized in the 1990s. Sieber argues that the scope of demonstrated and expected computer crimes today and in the future has also expanded far beyond economic crime, to cover attacks against national infrastructure and social wellbeing (Goodman and Brenner 2002).

In Europe, legal reforms have taken place in many countries since the 1970s, reflecting a change in the legal paradigm. The criminal codes of most of these countries have focused on the protection of tangible objects. However, the revolution of ICTs, which greatly depends on incorporeal values and information, in the latter part of the twentieth century, has predicated the development of new legislation which seeks these incorporeal values. The first step of this development in most European countries started in 1973 and addressed the protection of privacy, as a response to emerging vast capabilities for collecting, storing and transmitting data by computer (Siber et al. 1998). "Data protection legislations" were enacted and have been constantly revised and updated, protecting the citizens' right to privacy with administrative, civil, and penal regulations.

The second step involved the repression of computer-related economic crimes and started at the beginning of the 1980s (Ibid). It was precipitated by the inadequacy of existing traditional criminal provisions, which protected visible, tangible, and physical objects against traditional crimes, in the advent of cybercrime. These new legislations addressed the new capabilities of cybercrimes to violate traditional objects through new media, to protect intangible objects such as computer software (Ibid). Many European countries enacted new laws fighting computer-related economic crime (including unauthorized access to computer systems). Whilst some countries operate under the legal provisions enacted since the early 1980s, other countries are currently amending these provisions again to reflect new challenges to computer-related criminal law posed by the fast developing computer technology (Ibid).

In 1980s, a third series of additions to national laws were enacted. This wave was directed toward protecting the intellectual property in the realm of ICTs. The new legislation included copyright protection for computer software, penal copyright law and legal protection of topographies.

A fourth wave of reform legislation with respect to illegal and harmful contents began in a few countries in the 1980s, but are expanding rapidly since the triumphant rise of the internet in the mid-1990s. Legal amendments adapting traditional provisions on the dissemination of pornography, hate speech or defamation to computer-stored data were passed in the United Kingdom in 1994 and in Germany in 1997. Special provisions clarifying the responsibility of service and access providers on the internet were enacted in the United States of America in 1996. A final group of issues—discussed in particular in the 1990s—concerns the creation of requirements for and prohibitions of security measures. This field of law includes minimum obligations for security measures in the interest of privacy rights or in the general public interest. It also covers prohibitions of specific security measures in the interest of privacy rights or the effective prosecution of crimes, such as limitations of cryptography.

8.2.1 The Council of Europe Convention on Cybercrime

The 2001 Convention on Cybercrime ('Cybercrime Convention') of the Council of Europe is the most comprehensive international legislative effort to combat cybercrime to date (Koops 2011). It was signed in Budapest on November 23, 2001 by member states of the Council of Europe and by several non-member states, including Canada, Japan, South Africa, and the United States, that participated in its development. As of 5th June 2011 there were 16 signatory states not followed by ratifications and 31 countries which had ratified it and entered it into force, including the United States.[1] A further protocol on racist and xenophobic acts in cyberspace was signed on 28 January 2003, and entered into force on 1 March 2006.[2] Concurrent with both the convention and the protocol are explanatory reports. The Convention and its Explanatory Report have been adopted by the Committee of Ministers of the Council of Europe at its 109th Session (8 November 2001).

[1] Due to its Article 36, which contains the conditions for entry into force. It specifies that the Convention should first be ratified by five States, including three Member States of the Council of Europe. The Convention would then enter into force on the first day of the month following the expiration of a 3 month period after the fifth ratification. This condition was fulfilled with Lithuania's ratification on 18 March 2004, triggering the entry into force on 1 July 2004.

[2] Available at :http://conventions.coe.int/Treaty/en/Treaties/Html/189.htm.

8.2.1.1 Effectiveness and Impact of the Convention

A major premise of the treaty is that by fostering international cooperation, nations can tackle the problem of the borderless nature of cybercrime by enabling pursuit beyond the borders of a single nation. However, one needs to examine which countries fall under the auspices of the Cybercrime Treaty and which do not. This should be compared against the known sources of cybercrime to see how many nations have or have not been addressed. 27 EC nations have joined to date but only 12 have ratified. Outside the EU, the Convention is seen as Western dominated, both during its development and at the current time. Of the few non-EC nations that have acceded, only the US has ratified. On the other hand, the Convention is often held up as a model law, even for countries unwilling to accede because the treaty is seen as too Western, or too demanding of resources. While the United States is still the country from which the most Cybercrime attacks have originated, according to the most recent Internet Crime Complaint Center (IC3) report, other nations not covered by the treaty are: in second place, Canada; in fifth place, India; and the Philippines, in tenth place.

The key question for the success of the Cybercrime Convention is perhaps whether it can entice into membership those countries known to harbor the ring-leaders of organized cybercrime? In Europe, for example, the nation of Russia has neither signed nor ratified the treaty. Russia however represents the largest and most populated nation in all of Europe (Robel 2006). Russia has near an estimated 24 million people with internet access. According to the Russian Interior Ministry's Bureau for Counteracting High-Tech Crimes, internet crime in Russia has increased by ten times as much in the past 5 years (Ibid). This is partly attributed to large expansion of the internet throughout Russia. Indeed, Russian hackers have been blamed for everything from a number of computer viruses to orchestrated extortion schemes, and online trading protection money for averting the loss of websites (Ibid).

In relation to the United States, there are also neighbors in the Western Hemisphere that are not a part of the Cybercrime Treaty. The Treaty does claim Canada, as a major source of cyber-attacks.

Even where developing world and Eastern European countries have the political will to take a stance against cybercrime, it is often difficult to justify allocating resources for it, when the beneficiaries will not be that state's own citizens but those of other countries (Brown et al. 2009). Despite this, the ongoing success of the Cybercrime Convention can be seen at a micro, as well as macro level. Many countries are in the process of harmonizing their laws to meet Cybercrime Convention standards whether or not they plan to join (Ibid). In other regions such as the Arab states, there may be a preference to put together their own regional instruments rather than accede—but in most cases these are very similar to the Convention. It is thus arguably a very successful instrument for international harmonization (Ibid).

8.2.1.2 United States Ratification Serving as a Precedent

One possible aspect of US ratification that needs to be examined here is whether or not US ratification of the Convention may have an effect on the global community as a whole?

One must take into account the differing focus on legal and social values in different nations. One major issue is that the European Union is known for much stricter privacy laws than that of the United States. It especially has a concern in dealing with the United States' Department of Homeland Security. To illustrate just how strict the EU privacy laws are, they specify that an individual must be provided with information regarding who is processing their data, the purpose of its processing, who has received the data, a clear means to access and correct the data, and the source of the data (Roble 2006). This difference in approach to privacy can be seen in many instances. An example is when the US Administration required that the EU provide access to Passenger Name Records data (PNR) on Europeans flying to the United States.

On 16 December 2003, the European Commission announced details of the agreement reached with the United States on the transfer of PNR data to US authorities (House of Lords, 21st Report of Session 2006/2007). The Commission negotiators obtained from the United States a number of concessions on the amount of data to be sent and how the data would be handled. However, in 2004 the European Parliament had expressed strong reservations about how and what sort of data would be exchanged, and argued that such an agreement infringes EU citizens' privacy law. It passed a resolution asking the European Court of Justice to rule on whether the agreement violates European law. On May 30, 2006 the European Court of Justice (ECJ) annulled the decisions of the European Council and the European Commission which contained requirements to ensure an adequate level of data protection for the processing and transfer of Passenger Name Records from the EU to the United States Bureau of Customs and Border Protection (CBP) in the U.S.

In its judgment, the ECJ decided that the Commission decision could not be validly based on the EU Data Protection Directive 1995/46 because the processing and transfer of PNR data concerns "third pillar" activities (i.e., matters of public security, policy or judicial cooperation) which are excluded from the Directive's scope. The ECJ found that this was the case regardless of the fact that the PNR data is collected and processed by air carriers within the context of the sale of plane tickets. As it turns out, the deal between the EU and the US to transfer passenger reservation data from EU carriers to the US Department of Homeland Security was annulled as of September 30, 2006 by the European Court of Justice.

This example clearly illustrates the difficulties of the United States in coming to terms with the stricter laws of the EU. These concerns may also play a major part in preventing the ratification of the Cybercrime Treaty by the major nations of the EU. Without the support of these signatory nations, the Treaty may lack any true force, even with the involvement of the United States.

8.3 The American Response

Since various cybercrimes could be prosecuted under many federal statutes, the following discussion provides a brief survey of just some of the severe penalties that a cybercriminal could face under federal law in the U.S.

8.3.1 *Communication Interference Statute*

The communications interference statute listed in 18 U.S.C. section 1362 defines a number of acts under which an individual can be charged with a telecommunications related crime, including:

- Maliciously destroying property such as cable, system, or other means of communication that is operated or controlled by the United States;
- Maliciously destroying property such as cable, system, or other means of communication that is operated or controlled by the United States Military;
- Willfully interfering in the in the working or use of a communications line;
- Willfully obstructing or delaying communication transmission over a communications line;
- Conspiracy to commit any of the above listed acts.

Cybercriminals convicted under section 1362 are subject to fines and imprisonment of up to 10 years.

8.3.2 *The Stored Communications Act (SCA)*

The Stored Communications Act (SCA) is a law that was enacted by the United States Congress in 1986. It forms part of the Electronic Communications Privacy Act and is codified as 18 U.S.C. sections 2701–2712. The SCA addresses voluntary and compelled disclosure of "stored wire and electronic communications and transactional records" held by third-party internet service providers (ISPs).

Section 2701 of the SCA provides criminal penalties for anyone who "intentionally accesses, without authorization, a facility through which an electronic communication service is provided or … intentionally exceeds an authorization to access that facility; and thereby obtains, alters, or prevents authorized access to a wire or electronic communication while it is in electronic storage in such system." The SCA targets two types of online service: "electronic communication services" and "remote computing services." The statute defines an electronic communication service as "…any service which provides to users thereof the ability to send or receive wire or electronic communications." A remote computing service is defined as "the provision to the public of computer storage or processing services by means of an electronic communications system." Section 2703 of the SCA describes the conditions under which the government is able to compel an ISP to disclose "customer or subscriber" content and non-content information for each of these types of service:

- **Electronic communication service**. If an unopened email has been in storage for 180 days or less, the government must obtain a search warrant. There has been debate over the status of opened emails in storage for 180 days or less, which may fall into this category or the "remote computing service" category.
- **Remote computing service**. If a communication has been in storage for more than 180 days or is held "solely for the purpose of providing storage or computer processing services" the government can use a search warrant, or, alternatively, a subpoena or a "specific and articulable facts" court order (called a 2703(d) order) combined with prior notice to compel disclosure. Prior notice can be delayed for up to 90 days if it would jeopardize an investigation. Historically, an opened or downloaded email held for 180 days or less falls into this category, on the grounds that it is held "solely for the purpose of storage."

The Act was invoked in the 2010 *Robbins v. Lower Merion School District* case, where plaintiffs charged two suburban Philadelphia high schools of secretly spying on students by surreptitiously and remotely activating webcams embedded in school-issued laptops the students were using at home, thus violating their right to privacy. The schools admitted to secretly snapping over 66,000 webshots and screenshots, including webcam shots of students in their bedrooms.

8.3.3 Wiretap Act

The Wiretap Act regulates the collection of actual content of wire and electronic communications. Codified in 18 U.S.C. sections 2510–2522, the Wiretap Act was first passed as Title III of the Omnibus Crime Control and Safe Streets Act of 1968 and is generally known as "Title III". Prior to the 1986 amendment by Title I of the ECPA, it covered only wire and oral communications. Title I of the ECPA extended that coverage to electronic communications. The Wiretap Act broadly prohibits the intentional interception, use, or disclosure of wire and electronic communications unless a statutory exception applies. In general, these prohibitions bar third parties (including the government) from wiretapping telephones and installing electronic "sniffers" that read internet traffic. However, when authorized by the Justice Department and signed by a United States District Court or Court of Appeals judge, a wiretap order permits law enforcement to intercept communications for up to 30 days. 18 U.S.C. sections 2516–2518 impose several formidable requirements that must be satisfied before investigators can obtain a Title III order. Most importantly, the application for the order must show probable cause to believe that the interception will reveal evidence of a predicate felony offence listed in section 2516. 18 U.S.C. section 2518(3)(a), (b).

18 U.S.C. section 2510(12) defines “electronic communication” as: any transfer of signs, signals, writing, images, sounds, data, or intelligence of any nature, transmitted in whole or in part by a wire, radio, electromagnetic, photo-electronic or photo-optical system that affects interstate or foreign commerce, but does not include (A) any wire or oral communication; (B) any communication made through a tone-only paging device; (C) any communication from a tracking device…; or (D) electronic funds transfer information stored by a financial institution in a communications system used for the electronic storage and transfer of funds. As a rule, a communication is an electronic communication if it is neither carried by sound waves nor can fairly be characterized as one containing the human voice (carried in part by wire).

In United States v. Councilman, 418 F.3d 67 (1st Cir. 2005) (en banc), the First Circuit held that the term “electronic communication” includes transient electronic storage that is intrinsic to the communication process. Hence, the interception of an e-mail message in such storage is an offense under the Wiretap Act. U.S. v. Ropp, NO. CR 04-300-GAF (C.D. Cal. Oct. 8, 2004). It is also held that the interception of keyboard keystrokes by a key logger is not a violation of the Wiretap Act because such transmissions are not “electronic communications” within the meaning of the statute, which requires that the intercepted transmissions be “in interstate commerce.”

8.3.4 Wire Fraud Statute

Another criminal tool at the government’s disposal is the Wire Fraud statute. Wire fraud, in the United States Code, is any criminally fraudulent activity that has been determined to have involved electronic communications of any kind, at any phase of the event. The involvement of electronic communications adds to the severity of the penalty, so that it is greater than the penalty for fraud (otherwise identical except for the non-involvement of electronic communications). As in the case of mail fraud, the federal statute is often used as a basis for a separate, federal prosecution of what would otherwise have been a violation only of state law. The crime of wire fraud is codified at 18 U.S.C. section 1343, and reads as follows:

> Whoever, having devised or intending to devise any scheme or artifice to defraud, or for obtaining money or property by means of false or fraudulent pretenses, representations, or promises, transmits or causes to be transmitted by means of wire, radio, or television communication in interstate or foreign commerce, any writings, signs, signals, pictures, or sounds for the purpose of executing such scheme or artifice, shall be fined under this title or imprisoned not more than 20 years, or both. If the violation affects a financial institution, such person shall be fined not more than $1,000,000 or imprisoned not more than 30 years, or both.

8.3.5 Can-Spam Act

The CAN-SPAM Act of 2003, signed into law by President George W. Bush on December 16, 2003, establishes the United States' first national standards for the sending of commercial email and requires the Federal Trade Commission (FTC) to enforce its provisions. The acronym CAN-SPAM derives from the bill's full name: *C*ontrolling the *A*ssault of *N*on-*S*olicited *P*ornography *and M*arketing Act of 2003. This is also a play on the usual term for unsolicited email of this type, spam. The bill was sponsored in Congress by Senators Conrad Burns and Ron Wyden. The CAN-SPAM Act is occasionally referred to as the "You-Can-Spam" Act because while the bill does not explicitly legitimize email spam, it preempts laws that allowed for (among other things) easier prosecution and rights to private action. In particular, it does not require e-mailers to get permission before they send marketing messages. The Act also prevents states from enacting stronger anti-spam protections, and prohibits individuals who receive spam from suing spammers. The Act has been largely unenforced, despite a letter to the FTC from Senator Burns, who noted that "Enforcement is key regarding the CAN-SPAM legislation." The law required the FTC to report back to Congress within 24 months of the effectiveness of the Act. On December 20, 2005 the FTC reported that the volume of spam had begun to level off, and due to enhanced anti-spam technologies, less was reaching consumer inboxes. A significant decrease in sexually-explicit email was also reported. Later modifications changed the original CAN-SPAM Act of 2003 by (1) Adding a definition of the term "person"; (2) Modifying the term "sender"; (3) Clarifying that a sender may comply with the act by including a post office box or private mailbox and (4) Clarifying that to submit a valid opt-out request, a recipient cannot be required to pay a fee, provide information other than his or her email address and opt-out preferences, or take any other steps other than sending a reply email message or visiting a single page on an internet website.

8.3.6 Other Federal Statutes Related to Cyber Security

8.3.6.1 The Sarbanes Oxley Act

Although the Sarbanes-Oxley Act of 2002 ("SOX"), crafted in the wake of the Enron collapse, would appear to have little to do with cybersecurity, several SOX provisions impact upon IT professionals (Pinguelo and Muller 2011). Specifically, because of SOX's focus on the reliability of accounting/financial records and the implementation of effective control systems, it is important that companies employ an IT control system that is capable of detecting internal fraud (Ibid). While such insider fraud is difficult to detect because insiders frequently have intimate knowledge of the controls themselves, processes that provide for things like access control, detection of unusual account or access activity, [and] checks and balances for records relating to financial reporting may provide an early warning of such fraudulent activity (Ibid). Additionally, although SOX is targeted at publicly traded companies, privately held companies would be well served by adopting its reform measures, including putting in place more stringent internal control systems.

8.3.6.2 The Health Insurance Portability and Accountability Act

Because health industry payers and providers collect and maintain large volumes of protected health information, as well as other sensitive personal and financial data, and conducting many transactions electronically, they make attractive targets for identity thieves and other cybercriminals (Stamer 2005). In October 2010, the personal information of approximately 280,000 Medicaid members, including members' health plan identification numbers and some health records, was put at risk when Keystone Mercy Health Plan and AmeriHealth Mercy Health Plan reported the loss of a portable computer drive containing the said information. Accordingly, Title II of HIPAA, known as the Administrative Simplification (AS) provisions, requires the establishment of national standards for electronic health care transactions and national identifiers for providers, health insurance plans, and employers. The Administration Simplification provisions also address the security and privacy of health data. The standards are meant to improve the efficiency and effectiveness of the nation's health care system by encouraging the widespread use of electronic data interchange in the U.S. health care system.

8.3.6.3 The Fair and Accurate Credit Transaction Act of 2003

The Fair and Accurate Credit Transactions Act of 2003 is a United States federal law, passed by the United States Congress on November 22, 2003, as an amendment to the Fair Credit Reporting Act. The Act contains provisions to help reduce identity theft, such as the ability for individuals to place alerts on their credit histories if identity theft is suspected, or if deployed overseas in the military, thereby making fraudulent applications for credit more difficult. Further, it requires secure disposal of consumer information. Recently, the House and Senate passed measures to exempt lawyers, accountants, doctors and other health care professionals and service providers from the rigors of the Red Flag Rule.

8.3.6.4 The Gramm–Leach–Bliley Act

The Gramm–Leach–Bliley Act (GLB), also known as the Financial Services Modernization Act of 1999, is an act of the 106th United States Congress (1999–2001) that includes provisions dedicated to the protection of consumer financial information held by banks, securities firms, insurance companies and other financial institutions.

The Safeguards Rule requires financial institutions to develop a written information security plan that describes how the company is prepared for, and plans to continue to protect clients' nonpublic personal information. (The Safeguards Rule applies to information of any consumers, past or present, of the financial institution's products or services.) This plan must include:

- Denoting at least one employee to manage the safeguards,
- Constructing a thorough [risk management] on each department handling non-public information,
- Develop, monitor, and test a program to secure this information, and
- Change the safeguards as needed with the changes in how information is collected, stored, and used.

The Safeguards Rule forces financial institutions to take a closer look at how they manage private data and to do a risk analysis on their current processes. No process is perfect, implying that every financial institution has had to make some effort to comply with the GLB.

8.3.6.5 Children's Online Privacy Protection Act (COPPA)

COPPA and its associated regulations were put in place to put parents in control of what information is collected by commercial websites and online service providers on their children, under the age of 13, while they are online.

Website operators must use reasonable procedures to ensure they are dealing with the child's parent. These procedures may include:

- obtaining a signed form from the parent via postal mail or facsimile;
- accepting and verifying a credit card number;
- taking calls from parents on a toll-free telephone number staffed by trained personnel;
- email accompanied by digital signature;
- email accompanied by a PIN or password obtained through one of the verification methods above.

Operators who follow one of these procedures, acting in good faith to a request for parental access, are protected from liability under federal and state law for inadvertent disclosures of a child's information to someone who purports to be a parent.

8.3.6.6 Identity Theft and Assumption Deterrence Act of 1998 (ITADA)

The increase in crimes of identity theft led to the drafting of the Identity Theft and Assumption Deterrence Act of 1998 (FTC.gov). The statute now makes the possession of any "means of identification" to "knowingly transfer, possess, or use without lawful authority" a federal crime, alongside unlawful possession of identification documents. However, for federal jurisdiction to prosecute, the crime must include an "identification document" that either: (a) is purportedly issued by the United States, (b) is used or intended to defraud the United States, (c) is sent through the mail, or (d) is used in a manner that affects interstate or foreign commerce.

The Act also provides the Federal Trade Commission with authority to track the number of incidents and the dollar value of losses. Their figures relate mainly to consumer financial crimes and not the broader range of all identification-based crimes (Ibid). If charges are brought by state or local law enforcement agencies, different penalties apply, depending on the state.

Six Federal agencies conducted a joint task force to increase the ability to detect identity theft. Their joint recommendation on "red flag" guidelines is a set of requirements on financial institutions and other entities which furnish credit data to credit reporting services in order to develop written plans for detecting identity theft. The FTC has determined that most medical practices are considered creditors and are subject to requirements to develop a plan to prevent and respond to patient identity theft. These plans must be adopted by each organization's Board of Directors and monitored by senior executives.

The latest report from the FTC showed that ID theft was the number one complaint category in the Consumer Sentinel Network for the 2010 calendar year with 19 % of the overall complaints, followed by Debt Collection (11 %); Internet Services (5 %); Prizes, Sweepstakes and Lotteries (5 %); Shop-at-Home and Catalog Sales (4 %); Impostor Scams (4 %); Internet Auction (4 %); Foreign Money Offers and Counterfeit Check Scams (3 %); Telephone and Mobile Services (3 %); and Credit Cards (2 %). Florida was the state with the highest per capita rate of reported identity theft complaints, followed by Arizona and California.

Two states, California and Wisconsin, have created an Office of Privacy Protection to assist their citizens in avoiding and recovering from identity theft.

8.4 Technical Approach

As we move information security forward to face today's challenges, and those of tomorrow, we need to overcome a number of harmful forces.

Parker (1998), recommends several specific solutions for dealing with these challenges:

- We need to develop holistic new models and practitioners to convert information security from a folk art to a disciplined art.
- We need to create new laws to control the use of cryptography, clarify positions with regard to civil litigation and sting operations to catch computer criminals, and improve privacy through self-interest.

- Adding security to job performance appraisals will work miracles for raising management awareness and motivating individuals—managers and end users alike—to practice good security.
- Building unpredictability into our information systems will help to foil our adversaries.
- Making our computer applications similarly unpredictable, and advancing formal internal control models, can further foil attacks by our adversaries.
- We should replace the concept of risk assessment with the concept of due care to avoid negligence, which is more important and practical than avoiding risk, and accomplishes the same ultimate goal of protecting our information assets.
- We must divert youngsters from the hacker culture by making it socially unattractive and unacceptable.

8.5 The Future

There is no doubt that the future will bring new cybercrimes that we have not anticipated. Information security practitioners need to be perpetually alert to recognize them.

We won't see an end to new cybercrimoids with cute names any time in the near future. Because they gain, and hold, the public's attention, the media are likely to continue emphasizing this type of computer crime. The press would probably not have given so much attention to computer viruses if they were called *harmful replicating codes* (Parker 1998).

A stream of new technologies will be the target of criminals using ingenious new techniques that system designers naively, or in their haste, ignore. For example, *push* and *agent* technologies, in which unattended software works to find and deliver previously specified types of information and services, are ripe as subjects and objects of subversion and as tools for crime (Ibid). Other examples of new technology include speech recognition products, authentication of user identity services, three-dimensional displays for applications like spreadsheet computing, and commercial use of the internet. Embedded systems in smart homes, toys, appliances, clothing, sports equipment, and vehicles are also attractive for exploitation, thus making candidates as subjects of new cybercrimoids (Ibid).

The prospects, from the criminal's point of view, must be mouthwatering.

8.6 Summary

Cybercrime is a problem that is impossible to solve. It seems that our cyber-laws and law enforcement measures have always been one step behind the criminals. These difficulties remain the same with traditional crime. With respect to cyberspace, "[l]aw enforcement resources cannot keep pace with sophisticated cybercrime subcultures in anonymous offshore havens".[3] However, the nation's businesses must take whatever steps possible to fight cybercrime.

With the risk to national security and economy so great, and with carefully coordinated attacks from cybercriminals on the rise, law makers must act to pass an effective cyber-security package without delay.

And from the business perspective, although the defensive cyber-measures taken by larger companies will necessarily be complex and expensive, every business owner should focus on creating a level of cyber-awareness amongst staff to reduce the company's potential exposure to cyber-attacks. Simple measures such as moving information security forward, solving the hacker menace, dealing with the privacy problem and solving the cryptography information anarchy problem can go a long way towards making any company more "cyber-secure".

As noted by Renato Opice Blum "the reality is such that the profits from cybercrimes often surpass drug dealing and the question now lies on which prevention and punitive measures should be taken. At a minimum, awareness and education are necessary to keep up with the pace of these criminals".[4]

References

G. de Almeida, Legal rules and information security technical standards: possible approach for filling in the blanks of cybercrime legislation (2011), http://www.ssrn.com. Accessed 12 Aug 2011

I. Brown, L. Edwards, C. Marsden, in *Information Security and Cybercrime*, ed. by L. Edwards, C. Waelde. Law and the Internet (Hart, Oxford, 2009)

M. Goodman, S. Brenner, The emerging consensus on criminal conduct in cyberspace. Int. J. Law Inf. Technol. **10**(2), 139–223 (2002)

D. Griffith, The computer fraud and abuse act of 1986: a measured response to a growing problem. Vand. Law Rev. **43**, 453 (1990)

R. Hogge et al., Computer invasion of privacy under the Virginia Computers Criminal Act (2001), http://www.virginialaborlaw.com. Accessed 01 July 2011

B.-J. Koops, The internet and its opportunities for cybercrime (2011), http://www.ssrn.com. Accessed 03 July 2011

D. Parker, *Fighting Computer Crime: For Protecting Information* (Wiley, New York, 1998)

F. Pinguelo, B. Muller, Virtual crimes, real damages: a primer on cybercrimes in the united states and efforts to combat cybercriminals. Va. J. Law Technol. **16**(1), 116 (2011)

D. Robel, International cybercrime treaty: looking beyond ratification (2006), http://www.sans.iorg. Accessed 07 July 2011

D. Shinder, *Scene of the Cybercrime* (Syngress, Waltham, 2002)

[3] Fernando Pinguelo and Bradford Muller, op. cit. p. 188.

[4] Ibid.

U. Sieber et al., *The Legal Aspects of Computer Crime and Security: A Comparative Analysis with Suggestions for Future International Action* (Elsevier, New York, 1998)

C. Stamer, Cybercrime and identity theft: health information security beyond HIPAA. ABA Health eSource (2005), http://www.americanbar.org. Accessed 25 July 2011

Chapter 9
419 Scam: An Evaluation of Cybercrime and Criminal Code in Nigeria

9.1 The Meaning of 419 Scam

The term "419" is coined from Section 419 of the Nigerian criminal code (part of Chapter 38: obtaining property by false pretenses; cheating) dealing with fraud. Nowadays, the axiom "419" generally refers to a complex list of offences which in ordinary parlance are related to stealing, cheating, falsification, impersonation, counterfeiting, forgery and fraudulent representation of facts (Tive 2006, p. 3). The main difference between the "419" scam and stealing is that "false pretense" is the major element in the "419" scam. According to Section 23 of the advance fee fraud decree: "False pretense means a representation, whether deliberate or reckless, made by word, in writing or by conduct, of a matter of fact or law, either past or present, which representation is false in fact or law, and which the person making it knows to be false or does not believe to be true". Section 383, Sub-section 1 of the Nigerian Criminal Code states: "A person who fraudulently takes anything capable of being stolen, or fraudulently converts to his own use or to the use of any other person anything capable of being stolen, is said to steal that thing". It must be noted that falsification, impersonation, forgery and fraudulent representation of facts are all related tools that combine to either abet or facilitate the advance fee fraud (Ibid). Advance fee fraud did not start with the internet, although the internet enables the criminals to reach a greater number of potential victims more quickly and economically, and sometimes without being traced.

In the United Kingdom, the Audit Commission has conducted four triennial surveys of computer-related fraud based on a definition referring to: "any fraudulent behavior connected with computerization by which someone intends to gain financial advantage". Such a definition is capable of encompassing a vast range of activities some of which may have only the most tenuous connection with a computer. The Council of Europe, in its report on computer-related crime advocates the establishment of an offence consisting of: "The input, alteration, erasure or suppression of computer data or computer programs [sic], or other interference with

M. Chawki et al., *Cybercrime, Digital Forensics and Jurisdiction*,
Studies in Computational Intelligence 593, DOI 10.1007/978-3-319-15150-2_9

the course of data processing, that influences the result of data processing thereby causing economic loss or possessor loss of property of another person, or with the intent of procuring an unlawful economic gain for himself or for another person". However this definition is broad in scope. It would appear for example that the proposed offence would be committed by a person who wrongfully uses another party's cash dispensing card to withdraw funds from a bank account. Although there can be little doubt about the criminality of such conduct, the involvement of the computer is purely incidental. In most areas of traditional legal interests, the involvement of computer data does not cause specific legal problems. The respective legal provisions are formulated in terms of results and it is completely irrelevant if this result is achieved with the involvement of a computer or not.

The *Online Etymology Dictionary* associates 419 scam with a nineteenth century British slang "scamp", which means "cheater" or "swindler". Nevertheless, ever since 1963, when some scholars say the term "scam" made it into written literature for the first time, the central meaning has not changed—it is a trick, a ruse, a swindle, a racket (Igwe 2007, p. 6). Although 419 is broached as a global phenomenon, it did emerge from Nigeria and was the brainchild of a group of Nigerian nationals (Ibid). Therefore, a frontal battle against 419 ought not to overlook its place of origin and the numerous factors that gave birth to it. Many specialists in African studies claim, without substantiating details, that the game began in the 1980s with the Nigerian petroleum companies as major players. Some argue to the contrary, maintaining that 419 evolved from various types of tricks played since time out of mind, mostly in Igbo land in southeastern Nigeria (Ibid). Neither of these arguments is far from the truth, since both the oil industry and local intelligence have influenced the evolution of the scheme. The fact is that human beings tend to look for scapegoats all the time. In doing so, we frequently adopt limited views that undermine the idea that a single incident may be the result of a chain of combining and complementary forces and actions. A massive amount of advance fee fraud messages are sent out every day around the world, though many recipients ignore or discount their content. At the same time, a small percentage of all recipients respond to these messages and become victims who lose money or have their identities stolen at the hands of fraudsters (Graves and Holt 2007, p. 138).

9.2 The Operation of the Scheme

Advanced fee frauds commence with the receipt of an official-looking letter or email, usually purporting to be from the relative of a former senior government official, who, prior to their death, accrued a large amount of money which is currently being held in a bank account within the country from which the letter was being sent (Wall 2007, p. 90). The sender of the following typical 419 letter invites the recipients to assist with the removal of the money by channeling it through his or her bank account. In return for collaborating, the recipient is offered $12 million,

20 % of the 460 million to be transferred (Wall 2007, p. 91). Once recipients respond to the sender, an advanced fee is sought to pay for banking fees and currency exchange. As the victim becomes more embroiled in the scam and pays out money, it becomes harder to withdraw. Needless to say, the majority, if not all, of these invitations are bogus and are designed to defraud the respondents, sometimes for considerable amounts of money (Wall 2007, p. 92). Another technique employed by the scammers is to invite the victim to visit the scammers in their 'home' country to explain their situation in person and ask for money and assistance. This ploy is relatively uncommon, though it can lead the victim to be held hostage or killed (Graves and Holt 2007, p. 140). A final method requires the victim to provide the scammer with personal information, such as their name, address, employer, and bank account information. The initial request may be made under the guise of assuring the sender that the recipient is a sound and trustworthy associate (Holt et al. 2007, p. 140). However the information is surreptitiously used by the sender to drain the victim's accounts and engage in identity theft.

9.3 The Nature and Extent of the Problem

Advance fee fraud is a vexing threat and a major problem. It takes diverse forms and degrees, ranging from advancing sums of money, to murder (Tenfa 2006, p. 11). Furthermore, law enforcement officers often find it difficult to identify and apprehend cyber scammers. This may be due to the fact that perpetrators can use technology to conceal their identities and physical location, thereby frustrating law enforcement efforts to locate them (Chawki and Wahab 2006, p. 5). The traditional model of law enforcement assumes that the commission of an offence involves physical proximity between perpetrator and victim (Brenner 2004, p. 6). This assumption has shaped our approaches to criminal investigation and prosecution. Real-world criminal investigations focus on the crime scene as the best way to identify a perpetrator and link him to the crime (Chawki 2008, p. 13). However, in automated or cybercrime there may either be no crime scene or there may be many crime scenes, with shredded evidence of the crime is scattered throughout cyberspace. In this respect Dana van der Merwe (2008, p. 104) argues that:

> [T]he true problem of the information and communication era seems to be to decide exactly how much value should be attached to a given piece of information, especially when that information is stored electronically and digitally. The only field of law which advertises itself as a specialist in the area of verifying facts is the law of evidence. Unfortunately, like all other fields of law this field sometimes finds itself struggling to adapt to a new world in which paper is being phased out of general commercial transactions and to decreasing contact between human beings and the information needed to conduct business.

Accordingly, identifying an electronic crime scene can be a daunting task when the perpetrator may have routed his communications with the victim through computers in three or four countries, with obscure networks that are inaccessible to investigators (Chawki and Wahab 2006, p. 5). Very few estimates are available for advance fee fraud, and many of those published are more than a decade old. With regard to the United States, a number of conflicting figures have been published. Some examples include the following (Bocij 2006, p. 103):

- The FBI reported that the number of victims of advance fee fraud on average had tripled since 2001, as had total losses, growing from $17 million to $54 million.
- An estimate that has been quoted widely in the media claims that losses in the United States amount to $1 million to $54 million.
- In 2005, the average loss from 419 frauds was estimated at $6,937 by the National Consumers League. Other advance fee frauds resulted in an average loss of $1,426. However, an estimate from the FBI places the average loss from 419 frauds in 2005 at $3,000.

It is also difficult to obtain realistic estimates for other countries. However, the available information suggests that average losses are similar to those experienced in the United States. Some examples of losses experienced by other countries include (Bocij 2006, p. 103):

- In Canada, Phone Busters received 167 complaints over the period from January 2004 to September 2004. Victims lost a total of approximately $4.2 million.
- A 2004 estimate suggests that annual losses in South Africa are in the region of R100 million (approximately $16.5 million).
- In the United Kingdom, the National Criminal Intelligence Service (NCIS) reported that 150 people were defrauded for a total of £8.4 million, an average loss of £56,675, or almost $100,000. Total losses each year could be in the region of £150 million, or approximately $262 million.

9.4 Evaluation of the Current Situation in Nigeria

According to 2010 Internet Crime Report prepared by the National White Collar Crime Centre and the FBI, Nigeria currently ranks third in the world with 5.8 % of perpetrators of cybercrime (2010 Internet Crime Report, p. 11). Though the perpetrator percentage of 5.8 from Nigeria appears low, we can regard it as rather high considering that less than 10 % of the 150 million population of Nigeria use the internet (Delta State University Report). Cybercrime has a negative impact on Nigeria. It can be explained in the terms of the following statistics:

- Annual global loss of $1.5 billion in 2002.
- 6 % of global Internet spam in 2004.
- 15.5 % of total reported FBI fraud in 2001.
- Highest median loss of all FBI Internet fraud of $5,575.
- VeriSign, Inc., ranked Nigeria 3rd in total number of Internet fraud transactions, accounting for 4.81 % of global Internet fraud.

- American National Fraud Information Centre reported Nigerian money offers as the fastest growing online scam, up 900 % in 2001.

Nigerian Cybercrime has the potential to impact technology growth which is a key requirement for productivity improvement, and ultimately for socio-economic growth because:

- International financial institutions now view paper based Nigerian financial instruments with scepticism. Nigerian bank drafts and checks are not viable international financial instruments.
- Nigerian ISPs and email providers are already being black-listed in e-mail blocking blacklist systems across the internet.
- Some companies are blocking entire internet network segments and traffic that originate from Nigeria.
- Newer and more sophisticated technologies are emerging that will make it easier to discriminate and isolate Nigerian e-mail traffic.
- Key national infrastructure and information security assets are likely to be damaged by hostile and fraudulent unauthorized use.

Accordingly cybercrime has created an image nightmare for Nigeria. When one comes across phrases like 'Nigerian scam', the assumption that crosses one's mind is that all (or conservatively most) scam e-mails originate from Nigeria or Nigerians —though this is actually not the case (Adomi 2008, p. 720). Advance fee fraud has brought disrepute to Nigeria from all over the world. Essentially, Nigerians are treated with suspicious in business dealing. Consequently, the honest majority of Nigerians suffer as a result (Adomi 2008, p. 720).

9.5 419 Scam Combating Efforts in Nigeria

It has been argued that organized crime weakens the very foundation of democracy, as there can be no good governance without rule of law (Ngor, p. 172). This observation is quite apt for the situation in Nigeria. As the nation faces the challenges of nurturing a stable democracy, after many years of military dictatorship, organized crime poses a great threat to the survival of the country (Ngor, p. 172). Therefore, the Nigerian government has mapped out policies and strategies to deal decisively with crimes that are transnational in nature and scope.

9.5.1 *Legislative Approaches*

The Nigerian government has over the years enacted far-reaching laws aimed at checkmating transnational organized crime and punishing the perpetrators of these crimes. Under this sub-section we shall focus on the Criminal Code Act, Economic and Financial Crimes Commission Act 2004, Computer Security and Critical Information Infrastructure Protection Bill 2005 and Advance Fee Fraud and other Fraud Related Offences Act 2006.

9.5.1.1 Criminal Code Act

Advance fee fraud scam under the Nigerian Criminal Code Act, qualifies as a false pretense (Section 418), while a successful internet scam would amount to a felony under Section 419. This section provides as follows:

> "Any person who by any false pretense, and with intent to defraud, obtains from any other person anything capable of being stolen, or induces any other person to deliver to any person anything capable of being stolen, is guilty of a felony, and is liable to imprisonment for three years. If the thing is of the value of one thousand naira or upwards, he is liable to imprisonment for seven years. It is immaterial that the thing is obtained or its delivery is induced through the medium of a contract induced by the false pretense. The offender cannot be arrested without warrant unless found committing the offence".

Furthermore, a suspect could alternately be charged under Section 421 of the Criminal Code Act which provides as follows:

> "Any person who by means of any fraudulent trick or device obtains from any other person anything capable of being stolen, or induces any other person to deliver to any person anything capable of being stolen or to pay or deliver to any person any money or goods, or any greater sum of money or greater quantity of goods than he would have paid or delivered but for such trick or device, is guilty of a misdemeanor, and is liable to imprisonment for two years. A person found committing the offence may be arrested without warrant".

The Criminal Code is a British legacy which predates the internet era and understandably does not specifically address email scams (Oriola 2005, p. 240). Advance fee fraud methodology obviously falls within the remit of the act for the following reasons: first, there is a false pretense to the existence of non-existent money; second, a solicitation for financial help to get the fictitious money released; and third, the fraudulent retention of various fees paid to the scammers to release the phony millions of dollars. The scammers' modus operandi fits snugly the elements of the offence under Section 419 of the Criminal Code Act cited above, and have been used for years by the Nigerian law enforcement agencies for prosecuting alleged acquisition of property by false pretense (Oriola 2005, p. 240). However, the Criminal Code Act provisions on advance fee fraud are ill-suited for cyberspace criminal governance. Oriola argues that (2005, p. 241):

> Although Section 419 of the Criminal Code Act deems advance fee fraud a felony, the provision that an advance fee fraud suspect cannot be arrested without a warrant, unless found committing the offence, does not reflect the crime's presence or perpetration in cyberspace. Only in rare circumstances could a suspect be caught in the act because most of the scam emails are sent from Internet cafe´s. Aside from the fact that the country lacks the resources to police every known cyber cafe´, doing so could actually raise privacy or other rights issues (…) If found guilty, an advance fee fraudster is liable to a mere three years imprisonment or seven years if the value of stolen property exceeds 1000 naira. The

punishment, to say the least, is paltry relative to the enormity of the crime and unjust rewards that characteristically run into millions of dollars. Thirdly, in criminal trials, the State is the complainant, and there is hardly any compensation for victims of crime under the Nigerian criminal justice system. The victims could no doubt resort to civil court for remedies. However, the prospects for success for the plaintiff in the typical advance fee fraud case scenario are extremely slim. For instance, a contract to assist in the transfer from Nigeria of millions of dollars illegally to a foreign account, or to pay bribes to certain government officials to ensure release of such monies, or to facilitate advance fee payment for patently illegal activities, would be unenforceable. The plaintiff would be branded as a party to a culpable crime by the Nigerian courts.

9.5.1.2 Economic and Financial Crimes Commission Act 2004

The Economic and Financial Crimes Commission (Establishment) Act was adopted in June 2004. It repealed the Financial Crimes Commission Act of 2002 and establishes a Commission for Economic and Financial Crimes. Under this act, the Commission has the power to investigate all financial crimes relating to terrorism, money laundering, drug trafficking, etc. Sections 14–18 stipulate offences within the remit of the Act. This includes offences in relation to financial malpractices, offences in relation to terrorism, offences relating to false information and offences in relation to economic and financial crimes. The Act defines Economic and Financial Crimes as:

> "The non-violent criminal and illicit activity committed with the objectives of earning wealth illegally either individually or in a group or organized manner thereby violating existing legislation governing the economic activities of government and its administration and includes any form of fraud, narcotic drug trafficking, money laundering, embezzlement, bribery, looting and any form of corrupt malpractices, illegal arms deal, smuggling, human trafficking and child labor, illegal oil bunkering and illegal mining, tax evasion, foreign exchange malpractices including counterfeiting of currency, theft of intellectual property and piracy, open market abuse, dumping of toxic wastes and prohibited goods, etc.".

Although this definition does not refer directly to advance fee fraud or internet scam it could be argued that a direct reference to email frauds in the Economic and Financial Commission Act is superfluous and therefore unnecessary, since the Commission is already charged inter alia, with administering the Advance Fee Fraud and other Related Offences Act, which directly governs advance fee fraud in cyberspace (Oriola 2005, p. 244).

9.5.1.3 Money Laundering Prohibition Act 2004

Another related law on internet scam regulation in Nigeria is the Money Laundering (Prohibition) Act 2004. It makes provisions to prohibit the laundering of the proceeds of crime or an illegal act. Although advance fee fraud is not expressly mentioned in the Act, proceeds of the scam would appear covered under Section 14 (1)(a), which prohibits the concealing or disguising of the illicit origin of resources or property which are the proceeds of illicit drugs, narcotics or any other crime. The Act also implicates any person corporate or individual who aids or abet illicit disguise of criminal proceeds. Section 10 makes life more difficult for money launderers. Sub-section (1) places a duty on every financial institution to report within 7 days to the Economic and Financial Crimes Commission and the National Drug Law Enforcement Agency any single transaction or transfer that is in excess of ¼N1 m (or US$7,143) in the case of an individual or ¼N5 m (US$35,714) in the case of a body corporate (Chukwuemerie 2006, p. 178). Any other person may, under Sub-section (2), also give information on any such transaction or transfer. Under Sub-section (6) even if a transaction is below US$5,000, or equivalent in value, but the financial institution suspects or has reasonable grounds to suspect that the amount involved in the transaction is the proceed of a crime or an illegal act, it shall require identification of the customer. In the same way, if it appears that a customer may not be acting on his own account, the financial institution shall seek from him by all reasonable means information as to the true identity of the principal participant (Chukwuemerie 2006, p. 177). This enables authorities to monitor and detect suspicious cash transactions.

9.5.1.4 Computer Security and Critical Information Infrastructure Protection Bill 2005

In 2005, the Nigerian government adopted the Computer Security and Critical Information Infrastructure Protection Bill (known as the Cybercrime Bill). The Bill aims to "secure computer systems and networks and protect critical information infrastructure in Nigeria by prohibiting certain computer based activities", and to impose liability for global crimes committed over the internet. The Bill requires all service providers to record all traffic and subscriber information and to release this information to any law enforcement agency on the production of a warrant. Such information may only be used for legitimate purposes as determined by a court of competent jurisdiction, or other lawful authority. The Bill does not provide independent monitoring of the law enforcement agencies carrying out the provisions, nor does the Bill define "law enforcement agency" or "lawful authority." Finally the Bill does not distinguish between serious offences and emergencies or minor misdemeanors. As a result it may conflict with Article 37 of Nigeria's Constitution, which guarantees the privacy of citizens including their homes and telephone conversations, absent a threat on national security, public health, morality, or the safety of others.

9.5.1.5 AFF and Other Fraud Related Offences Act 2006

Another relevant legislative measure in the fight against advance fee fraud on the internet is the Advance Fee Fraud and other Fraud Related Offences Act 2006. This is a replacement of an act of the same title passed in 1995. The Act prescribes, among others, ways to combat cybercrime and other related online frauds. The Act provides for a general offence of fraud with several ways of committing it, which are by obtaining property by false pretense, use of premises, fraudulent invitation, laundering of funds obtained through unlawful activity, conspiracy, aiding, etc. Section 2 makes it an offence to commit fraud by false representation. Sub-sections (2)(a) and (2)(b) make clear that the representation must be made with intent to defraud. Section 3 makes it an offence if a person who is the occupier or is concerned in the management of any premises, causes or knowingly permits the premises to be used for any purpose which constitutes an offence under this act. This section provides that the sentence for this offence is the imprisonment for a term of not less more than 15 years and not less than 5 years without the option of a fine.

Section 4 refers to the case where a person, by false pretense and with the intent to defraud any other person, invites or otherwise induces that person or any other person to visit Nigeria for any purpose connected with the commission of an offence under this Act. The sentence for this offence is the imprisonment for a term not more than 20 years and not less than 7 years without the option of a fine.

According to Section 7, a person who conducts or attempts to conduct a financial transaction which involves the proceeds of a specified unlawful activity with the intent to promote the carrying on of a specified unlawful activity; or where the transaction is designed to conceal or disguise the nature, the location, the source, the ownership or the control of the proceeds of a specified unlawful activity, is liable on conviction to a fine of Nl million and in the case of a director, secretary or other officer of the financial institution or corporate body or any other person, to imprisonment for a term, not more than 10 years and not less than 5 years.

However, while in previous laws the onus was on the government to carry out surveillance on such crimes and alleged criminals, the new law vests this responsibility on industry players, including ISPs and cybercafé operators, among others. While the Economic and Financial Crimes Commission (EFCC) becomes the subsector regulator, the Act prescribes that henceforth, any user of internet services shall no longer be accepted as anonymous. Through what has been prescribed as due care measure, cybercafés operators and ISPs will henceforth monitor the use of their systems and keep a record of transactions of users (Adomi 2008, p. 290). These details include, but are not limited to, photographs of users, their home address, telephone, email address, etc. So far, over 20 cybercafés have been raided by the EFCC as of August 7, 2007. The operators appear set to comply with the law by notifying users of the relevant portion of the law, corporate user policy, firewall recommendation, protection procedure, indemnity, right of disclosure, and so forth (Adomi 2008, p. 290).

9.5.2 Administrative Measures

Administrative measures chiefly involve the setting-up of special bodies by the Nigerian government to combat advance fee fraud. Equally important, however, are the technical measures which these bodies then take to prevent and/or prosecute this activity, which will also be examined below. It must be emphasized that many European countries have established special computer units to take specific measures against cybercrime. The following are some examples of these bodies.

9.5.2.1 The Economic and Financial Crimes Commission (EFCC)

The Economic and Financial Crimes Commission (EFCC) is a Nigerian law enforcement agency that investigates financial crimes such as advance fee fraud (419 fraud) and money laundering. The EFCC was established in 2003, partially in response to pressure from the Financial Action Task Force on Money Laundering (FATF), which named Nigeria as one of 23 countries non-cooperative in the international community's efforts to fight money laundering. EFCC is an inter-agency commission, comprising a 22-member board drawn from all the Nigerian Law Enforcement Agencies (LEAs) and Regulators. The commission is empowered to investigate, prevent and prosecute offenders who engage in 'money laundering, embezzlement, bribery, looting and any form of corrupt practices, illegal arms deal, smuggling, human trafficking, and child labor, illegal oil bunkering, illegal mining, tax evasion, foreign exchange malpractices including counterfeiting of currency, theft of intellectual property and piracy, open market abuse, dumping of toxic wastes, and prohibited goods'.

The commission is also responsible for identifying, tracing, freezing, confiscating, or seizing proceeds derived from terrorist activities. EFCC is also host to the Nigerian Financial Intelligence Unit (NFIU), vested with their responsibility of collecting suspicious transactions reports (STRs) from financial and designated non-financial institutions, analyzing and disseminating them to all relevant government agencies and other FIUs all over the world. In addition to any other law relating to economic and financial crimes, including the criminal and penal codes, EFCC is empowered to enforce all anti-corruption and anti-money laundering laws. Punishment prescribed in the EFCC Establishment Act range from a combination of fine payment, forfeiture of assets and up to 5 years imprisonment, depending on the nature and gravity of the offence. Conviction for terrorist financing and terrorist activities attracts life imprisonment. It must be mentioned that EFCC has an excellent working relationship with major law enforcement agencies all over the world. These include the INTERPOL, the UK Metropolitan Police, FBI, Canadian Mounted Police, the Scorpions of South Africa, etc.

9.5.2.2 Nigerian Financial Intelligence Unit (NFIU)

In 2005, the EFCC established the Nigerian Financial Intelligence Unit (NFIU). The NFIU draws its powers from the Money Laundering (Prohibition) Act of 2004 and the Economic and Financial Crimes Commission Act of 2004. It is the central agency for the collection, analysis and dissemination of information on money laundering and terrorism financing. All financial institutions and designated non-financial institutions are required by law to furnish the NFIU with details of their financial transactions. Provisions have been included to give the NFIU power to receive suspicious transaction reports made by financial institutions and non-designated financial institutions, as well as to receive reports involving the transfer to or from a foreign country of funds or securities exceeding $10,000 in value (International Narcotics Control Strategy Report 2006, p. 2). The NFIU is a significant component of the EFCC. It complements the EFCC's directorate of investigations but does not carry out its own investigations. It is staffed with competent officials, many with degrees in accounting and law (International Narcotics Control Strategy Report 2006, p. 2). Moreover, the NFIU is playing a pivotal role in receiving and analyzing STRs. As a result, banks have improved their responsiveness to forwarding records to the NFIU (International Narcotics Control Strategy Report 2006, p. 2). Under the EFCC Act, whistle-blowers are protected. Nigeria has no secrecy laws that prevent the disclosure of client and ownership information by domestic financial services companies to bank regulatory and law enforcement authorities (International Narcotics Control Strategy Report 2006, p. 2). The NFIU has access to records and databanks of all government and financial institutions, and it has entered into memorandums of understandings (MOUs) on information sharing with several other financial intelligence centers. The establishment of the NFIU is part of Nigeria's efforts toward removal from the NCCT list (International Narcotics Control Strategy Report 2006, p. 3).

9.5.2.3 Nigerian Cybercrime Working Group (NCWG)

The NCWG is an inter-agency body comprising law enforcement, intelligence, security as well as ICT agencies of government and key private sector ICT organizations (NCWG website 2008). It was established by the Federal Executive Council (FEC) on the recommendation of the President of Nigeria on March 31, 2004. The group was created to deliberate on and propose ways of tackling the malaise of internet 419 in Nigeria. This includes (NCWG website 2008):

- Educating Nigerians on cybercrime and cybersecurity;
- Undertaking international awareness programs for the purpose of informing the world of Nigeria's strict policy on cybercrime and to draw global attention to the steps taken by the government to rid the country of internet 419 in particular, and all forms of cybercrime;
- Providing legal and technical assistance to the National Assembly on Cybercrime and Cybersecurity in order to promote general understanding of the subject matter amongst the legislators;
- Carrying out institutional consensus building and conflict resolutions amongst law enforcement, intelligence and security agencies in Nigeria for the purpose of easing any jurisdictional or territorial conflicts or concerns of duties overlap;

- Reviewing, in conjunction with the Office of the Attorney General of the Federation, all multilateral and bilateral treaties between Nigeria and the rest of the world on cross-border law enforcement known as Mutual Legal Assistance Treaties (MLAT), for the purpose of amending the operative legal framework to enable Nigeria secure from, as well as render, extra-jurisdictional assistance to its MLAT partners in respect to cybercrime.

9.5.3 Technical Measures

Criminals are often quicker to exploit new technologies than law enforcers who, to some extent, always seem "behind the game". In order to salvage Nigeria from the negative consequences of cybercrime, the government has been making frantic efforts to ensure that this malaise is "nipped in the bud". These efforts are discussed below:

9.5.3.1 Regulation of Cybercafés

The "cybercafé", also known as the "internet café" or "PC café", is a place where public access internet services are provided by entrepreneurs for a fee (Adomi 2007). While in the US and Western Europe, the term "cybercafé" refers often to true cafes offering both internet access and beverages, in Nigeria and other parts of Africa, cybercafés can refer to places offering public internet access in places like restaurants or hostels, as well as locations that are wholly set aside for public access internet services (Adomi 2007). Certain cybercafés in Nigeria even offer overnight browsing, a special internet service available from 10.00 p.m. to 6.00 a.m. This service allows users who have a lot to obtain from the net to do so at a minimal cost (Adomi 2007).

Though overnight browsing is very important and useful to cybercafé users, it was banned by the EFCC and the Association of Cybercafé and Telecentres Owners (ATCON) in Nigeria (Adomi 2008).The ban is coming on the heels of several attempts by the EFCC to arrest the trend through raids, arrests, and cautions of cybercafés and cyber criminals, as a result of the constant embarrassment posed to the Nigerian Federal Government by their nefarious activities. Some Nigerian fraudsters have perfected the act of using the internet via cybercafés as their criminal platform to dupe unsuspecting citizens across the globe (Adomi 2008). This ban on night browsing is likely to negatively affect clients who use the cafes for academic and other useful and positive purposes in night browsing sessions. Other decisions of EFCC and ATCON reached to combat cybercrime include (Adomi 2008):

- Undertaking international awareness programs for the purpose of informing the world of Nigeria's strict policy on cybercrime and to draw global attention to the steps taken by the government to rid the country of internet 419 in particular and all forms of cybercrimes;
- That each sector of the telecom industry, namely the global system for mobile communication operators, private telecom operators and cybercafés should come up with a due care document that would be a standard guide and proffer measures for the effective policing of cybercrime in Nigeria;
- That all cybercafés must be registered with the Corporate Affairs Commission, NCC and EFCC;
- That cybercafés will now be run on membership basis instead of pay-as-you-go;
- All cybercafés must install acceptable hardware surveillance;
- The architecture of cybercafés must be done such that all computers are exposed;
- ATCON members must subscribe to registered and licensed ISP in the country;
- Each cybercafé is expected to be a watchdog to others, as they have been detailed to have direct access to EFCC (Adomi 2008)

9.5.3.2 Government Partnership with Microsoft

The government of Nigeria and Microsoft Corp signed a Memorandum of Understanding defining a framework for cooperation between Microsoft and the Economic and Financial Crimes Commission (EFCC) of Nigeria with the aim of identifying and prosecuting cyber criminals, creating a safe legal environment and thereby restoring hundreds of millions of dollars in cost investment. This agreement is the first of its kind between Microsoft and an African government and will give the EFCC access to Microsoft technical expertise information for successful enforcement. The memorandum combats issues such as spam, financial scam, phishing, spyware, viruses, worms, malicious code launches and counterfeiting. Microsoft is expected to instruct Nigerian investigators on techniques of extracting useful information from PCs compromised by botnet attacks, how to monitor computer network to detect such attacks, and how to identify the people behind them. Microsoft will also provide leads on spam emanating from Nigeria, enabling the authorities to pursue investigations more quickly and successfully (Adomi 2008). Microsoft is known for conducting a worldwide analysis of spam sent to e-mail accounts that it establishes and monitors for this purpose (Adomi 2008).

9.5.3.3 ADNET

ADNET is a computer network with powerful capabilities for the storage and retrieval of data concerning Nigerian crime. ADNET is a secure system and can be accessed through dedicated ADNET terminals in the task force cities. In conjunction with the NCI working group, an outside private contractor trains and provides support to investigators working Nigerian crime cases. ADNET terminals are also located in Lagos, Nigeria and Accra, Ghana, so that data can be accessed close to the source of much of the Nigerian criminal activity. Several federal law enforcement agencies contribute and access ADNET data. In the last 2 years the number of records in the NCI database has increased dramatically, making the network a potentially highly valuable resource to law enforcement. Some of this data consists of information collected from prior criminal investigations, including aliases used by persons involved in Nigerian criminal activities.

9.6 The Quest for Legislative Harmonisation

Considerable differences still exist in legislative responses to advance fee fraud. Despite common legal histories, there are differences in the definition of crimes, the penalties applicable to forms of these crimes and the extent of criminalization. Although these differences reflect the position of criminal law in general in each country, there is increasing recognition that legislative differences can impede effective law enforcement with the international nature of much advance fee fraud. In this regard, governments must provide for:

- Effective criminalization of cyber–offences. The legislation of different countries should be as harmonized as possible.
- Conditions facilitating direct cooperation between state institutions, as well as between state institutions and the private sector.
- Investigative procedures and institutional capacities which allow criminal justice agencies to cope with advance fee fraud.

The "Budapest" Convention on Cybercrime (ETS 185) of the COE helps countries to respond to these needs. It is thus imperative that countries ratify this convention since it is an effective tool for counter action. What is needed is close co-operation and co-ordination and also harmonised arrangements for effective prosecution of transnational criminals. As noted in the 2003 Council of Europe report, effective law enforcement requires 'sufficient resources to finance law enforcement units trained and dedicated to fight cybercrimes'. The development of methods and tools to provide information and assistance for victims of advance fee fraud by NGOs and other agencies active in the field of cybercrime must be supported.

9.7 Future Trends

Further research is needed on other unexplored areas of internet based fraud, e.g. matrimonial web sites, where the internet is more likely to be used to attract victims, or facilitate transactions with them. Prospects for cooperation with computer experts, and businesses which operate internet gateways should be explored. There are already examples of codes of conduct, which have been introduced online to tackle the problem of cybercrime and advance fee fraud. A self-regulation system should also be introduced. Self-regulation is based on three key elements: first, involvement of all interested parties (government, user associations, and access providers) in producing new strategies; secondly, implementation of these strategies by the party concerned; and thirdly, evaluation of the measures taken. Self-regulation can be backed by clear legal regulation, also known as 'co-regulation'. A co-regulatory system is one in which the public authorities accept that protection of the society can be left to self-regulatory schemes, but reserve the right to intervene, if self-regulation fails to work.

For the purpose of prosecuting advance fee fraud scammers, we should use: local, regional and national co-ordination and information sharing mechanisms; national liaison officers posted overseas or links with liaison officer networks; Europol and its Liaison Bureaux; Interpol's National Contact Bureaux; Eurojust; and direct bi-lateral contacts. Channels that already exist for other purposes should be activated and adapted. Finally, we need to make laws more effective by improving the quality of criminal codes and increasing penalties to match the seriousness of loss. Laws will be effective against scammers who are deterred by criminal law and frightened by the prospect of incarceration; however, there will always be scammers who are motivated to engage in online fraud to overcome these laws and the efforts of the criminal justice community.

9.8 Summary

Cyberspace is still very largely terra incognita, and that leaves plenty of scope for criminal activity. It should be noted that technology is truly a double-edged sword that has transformed the classical and traditional forms of criminal behavior. The proliferation of ICTs and progressive development in digital transactions and communications have created new opportunities and opened up new windows which have resulted in the emergence of new forms of criminal behaviour and cybercrime. On such a basis, advance fee fraud ranks amongst the most important and virulent forms of cybercrime; not only due to its adverse impact on the development of cyberspace but also due to the diversity of means and methods that could be utilized in committing this crime as well as the inherent risk of using advance fee fraud as a leeway and instrument to commit other crimes using the stolen identities of victims. Furthermore, advance fee fraud could have a devastating impact on the financial security and credit scoring of victims. Being aware of the potential and actual risks associated with this serious exploitation of ICTs, the author has, throughout this chapter, attempted to provide a comprehensive overview of the fundamental issues and potential solutions pertinent to this form of criminal behavior. Largely, it is submitted that advance fee fraud should be subject to a global principle of public policy that aims at combating and preventing this form of organized crime through raising global awareness, increasing literacy rates, coordinating legislative efforts on national, regional and global levels, establishing a high level global network of cooperation between national, regional, and international enforcement agencies and police forces.

Acknowledgement An earlier version of this chapter was published in JILT 2009 (1). The author is grateful to Professor Abdul Paliwala.

References

E. Adomi, Overnight internet browsing among cybercafé users in Abraka, Nigeria. J. Community Inf. **3**(2), (2007)
E. Adomi, Combating cybercrime in Nigeria. Electron. Libr. **26**(5), (2008)
P. Bocij, *The Dark Side of the Internet: Protecting Yourself and Your Family from Online Criminals* (Praeger Publishers, Westport, 2006)
S. Brenner, Cybercrime metrics: old wine, new bottles. Va. J. Law Technol. **9**(6), (2004)
M. Chawki, *Combattre la Cybercriminalité* (Editions de Saint-Amans, Perpignan, 2008)
M. Chawki, M. Wahab, Identity theft in cyberspace: issues and solutions. LexElectrnica **11**(1), (2006)
A. Chukwuemerie, Nigeria's Money Laundering (Prohibition) Act 2004: a tighter noose. J. Money Laundering Control **9**(2), (2006)
D. Graves, T. Holt, A qualitative analysis of advance fee fraud email schemes. Int. J. Cybercriminol. **1**(1), (2007)
C. Igwe, *Taking Back Nigeria from 419: What to do about the Worldwide Email Scam Advance Fee Fraud* (Bloomington, iUniverse, 2007)
T. Oriola, Advance fee fraud on the internet: Nigeria's regulatory response. Comput. Law Secur. Rev. **21**(3), (2005)
D. Tenfa, *Advance Fee Fraud* (University of South Africa, Pretoria, 2006)
C. Tive, *419 Scam: Exploits of the Nigerian Con Man* (Bloomington, iUniverse, 2006)
D. van der Merwe et al., *Information and Communications Technology Law* (Durban, Lexis Nexis, 2008)
D. Wall, *Cybercrime: The Transformation of Crime in the Information Age* (Polity Publishers, Cambridge, 2007)

Appendix

Cases

Director of Public Prosecutions v Lennon (2006) EWHC 1201
DPP v Bignell (1998) 1 Cr. App R.1
R. v. McLaughlin (1980) 2 S.C.R. 331
R v Tomsett (1985) Crim LR 369
US v. Alfonso Flores Ramirez, CR 00-60043-01-HO (D. Or. Aug. 30, 2000)
US v Alston 609 F 2d (DC Cir 1979), cert denied 445 US 918 (1980)
US v. Angel Sanchez, CR 00-60080-01-HO (D.Or. Aug. 31, 2000)
U.S. v. Baker, 890 F. Supp. 1375 (1995)
US v. Bosanac, no. 99CR3387IEG
US v. Cleotilde Fregoso Rios, CR 00-60035-01-HO (D.Or. Nov. 7, 2000)
US v Gallant 570 F Supp 303 (SDNY1983)
US v. Javier Hernandez Lopez, CR 00-60038-01-HO (D.Or. Aug. 10, 2000)
US v. Jose Arevalo Sanchez, CR 00-60040-01-HO (D. Or. Nov. 21, 2000)
US v. Jose Manuel Acevez Diaz, CR 00-60038-01-HO (D. Or. Aug. 10, 2000)
US v Kelly 507 F Supp 495, 498 n6 (ED Pa 1981)
US v. Lamar Christian, CR 00-3-1 (D. Del. Aug. 9, 2000)
US v. Maria Mersedes Calderon, CR 00-60046-01-HO (D.Or. May 10, 2000)
US v. Mark L. Simons, 206 F. 3d 392 (4th Cir., Feb. 28, 2000)
US. v. Pedro Amaral Avila, CR 00-60044-01-HO (D.Or. Nov. 7, 2000)
US. Ranulfo Salgado, CR 00-60039-01-HO (D.Or. Jan. 18, 2001)
US v. Ronald Nevison Stevens, CR 00-3-2
US v Seidlitz, 589 F. 2d 152 (4th Cir. 1978)
US v. Victor Manuel Carrillo, CR 00-60045-01-HO (D. Or. Oct. 24, 2000)
US v. Wahl, CR00-285P 13

M. Chawki et al., *Cybercrime, Digital Forensics and Jurisdiction*,
Studies in Computational Intelligence 593, DOI 10.1007/978-3-319-15150-2

Printed in the United States
By Bookmasters